Animal Cell Culture

NIPA® GENX ELECTRONIC RESOURCES & SOLUTIONS P. LTD.
New Delhi-110 034

About the Authors

Dr. Ratan Kumar Choudhary is an Assistant Professor at the College of Animal Biotechnology, Guru Angad Dev Veterinary and Animal Sciences University, Ludhiana, Punjab, India. Ratan received his terminal degree (Ph.D.) from the University of Maryland, College Park, USA. He has more than six years of experience at the United States Department of Agriculture (USDA) facility. He received two post-doctoral trainings- from the University of Kentucky and the University of Vermont, USA. He works in stem cell biology, developing regenerative medicine for wound healing, cancer, and mastitis. The other area of his research interests is manipulating mammary stem cells for lactation persistency. The long-term goal of his research is to develop stem cell therapies for pets and utilize stem cells as 'drug' in bovine mastitis. He published three books, more than 60 research papers, ten book chapters and contributed significantly by making more than 60 presentations of his research at national and international levels. He is the recipient of a prestigious international award (Cochran Fellow 2023) from the USDA. In addition, he received the Best Teacher of the College (2023), Best Researcher of the University (2023), and was selected as a Member of the National Academy of Veterinary Sciences.

Dr Manu M is a veterinary virologist currently posted as an Assistant Professor in the Department of Microbial and Environmental Biotechnology of College of Animal Biotechnology, Guru Angad Dev Veterinary and Animal Sciences University, Ludhiana, Punjab, India. He has completed his graduation from Kerala Veterinary and Animal Sciences University and post-graduation (MVSc) and PhD in the discipline of Veterinary Virology from ICAR-Indian Veterinary Research Institute. His areas of expertise include viral vaccines, sero-diagnostics and molecular epidemiology. He has gained much acclaim for the development of indigenous live attenuated cell culture Classical swine fever (CSF) vaccine, an alternate potency test for determination of PD_{50} of CSF vaccines and a competitive ELISA to detect CSF antibodies during his PhD research. He has published more than 15 research articles and several reviews in peer-reviewed national and international journals. He is the recipient of several national awards and has been associated with societies of national and international repute *vis*. World Society for Virology, Association of Microbiologists of India and Indian Virological Society.

Prof. Yashpal Singh Malik is currently the Dean of the College of Animal Biotechnology, Guru Angad Dev Veterinary and Animal Sciences University, Ludhiana, India. He is a recipient of the prestigious position "ICAR National Fellow" at the ICAR-Indian Veterinary Research Institute. His areas of expertise are viral disease epidemiology, microbial biodiversity, host-virus interactions and pathogen-diagnostics. He has pursued advanced studies in molecular virology at the University of Minnesota, USA; University of Ottawa, Ontario, Canada; and Wuhan Institute of Virology, Wuhan, China. He is the recipient of several prestigious national, state and academy awards and honors, including the ICAR-Jawaharlal Nehru Award. He has authored 9 books, 62 book chapters, and over 255 research and review articles. Prof Malik has been associated with societies of international repute, like, the Secretary General of the World Society for Virology (USA) and at national level serving as Secretary General of Indian Virological Society. Being a member in "One Health group in Federation of Asian Veterinary Association" (FAVA) for 2021-2025, he is the Indian flag bearer on the international forum. Prof Malik is the Editor-in-Chief of the Journal of Immunology Immunopathology. His h-index is 56 with over 13000 citations. He has been awarded a prestigious Fellowship by the National Academy of Agricultural Sciences, Academy of Microbiological Sciences, National Academy of Veterinary Sciences and National Academy of Dairy Sciences.

Animal Cell Culture
Volume 01: The Protocol Series

Ratan Kumar Choudhary
Assistant Professor
College of Animal Biotechnology
Guru Angad Dev Veterinary and Animal Sciences University
Ludhiana, Punjab, India

Manu M
Assistant Professor
Department of Microbial and Environmental Biotechnology
College of Animal Biotechnology
Guru Angad Dev Veterinary and Animal Sciences University
Ludhiana, Punjab, India

Yashpal Singh Malik
Dean
College of Animal Biotechnology
Guru Angad Dev Veterinary and Animal Sciences University
Ludhiana, Punjab, India

NIPA® GENX ELECTRONIC RESOURCES & SOLUTIONS P. LTD.
New Delhi-110 034

NIPA® GENX ELECTRONIC RESOURCES & SOLUTIONS P. LTD.

101,103, Vikas Surya Plaza, CU Block
L.S.C.Market, Pitam Pura, New Delhi-110 034
Ph : +91 11 27341616, 27341717, 27341718
E-mail: newindiapublishingagency@gmail.com
www: www.nipabooks.com

For customer assistance, please contact
Phone: + 91-11-27 34 17 17
Fax: + 91-11-27 34 16 16

Print ISBN: 978-93-58879-70-4

ebook ISBN: 978-93-58877-72-4

Composed and Designed by NIPA®.

Preface

Animal cell culture is a common laboratory technique for maintaining live cells separated from their original tissue (primary cells) or immortal cells (cell lines). The technique of cell culture was discovered in the 19th century after the discovery of the microscope. Presently, the culture of cells is one of the most important tools of life sciences research and is being used by investigators to understand advanced understanding of cell growth and differentiation and mechanisms of normal/abnormal functioning of various cells. In addition to that, cell culture is employed in the generation of vaccines, the production of pharmaceutical proteins, and the safety testing of drugs.

The Faculty of the College of Animal Biotechnology authored or adapted the protocol contained in this manual for the purpose of creating basic information resource for students beginning studies in the field of Biotechnology to the course "Principles and Procedures of Animal Cell Culture (Biotech 421). This manual may be useful for some other undergraduate program courses in animal biotechnology.

Starting with Experiment No. 1, you will delve into the fundamental principles of laboratory layout, safety protocols, and guidelines crucial for establishing and maintaining cell culture facilities. Subsequent experiments cover various topics, including basic equipment setup, sterilization techniques, media preparation, and isolation of primary cell cultures. Moreover, this manual incorporates emerging trends in cell culture, including cryopreservation methods, virus-mediated cell line development, and virus-neutralization assays, reflecting the evolving landscape of biomedical research.

Whether you're conducting experiments in academic, industrial, or clinical settings, this manual equips you with the knowledge and skills necessary to conduct reproducible cell culture experiments. We encourage you to explore, experiment, and discover the fascinating world of cell biology through hands-on laboratory experiences outlined in this manual.

Happy experimenting!

Authors

Contents

1

Layout, Safety Protocols and Guidelines in Cell Culture Facilities

The tissue/cell culture lab has the requirement of maintaining an aseptic environment. There are several aspects to the design of good tissue culture facilities. Ideally, cell culture work should be conducted in a single-use facility. Therefore, cell culture lab unit should be separated from the rest of the lab area. It can be either separate lab or a closed partition in a big lab. All new material should be handled as 'quarantine' material in a separate area, and materials known to be free of contaminants should only be allowed inside the culture room. In addition, the work surfaces should be thoroughly cleaned and disinfected between activities.

For most of the work involving non-infectious primary culture and cell lines, the laboratory should be designated to at least BSL2 facility. The location of the Biosafety cabinet is the most important point to consider. When operated correctly, the cabinet will provide a clean working environment, whilst protecting the operator from aerosols. Environmental monitoring with Tryptone Soy Broth agar settle plates inside the biosafety cabinet is a good indicator of the cleanliness of a cabinet. There should be no growth of bacteria and fungi after 3-5 days of incubation.

A sample layout of a cell culture laboratory is given below (Figure 1&2). Description of items (from clockwise direction) is provided with important considerations.

1. Main door- There should be one entry point with an asymmetrical double door.
2. Inverted microscope and cell counting machine or facility of counting cells using a hemocytometer should be adjacent to the laminar flow.
3. Laminar flow or biosafety cabinet should be placed away from direct contact with circulating air condition (A/C). It should be regularly inspected at every year for maintenance and proper functioning.
4. Trash cans- for normal, dry and uncontaminated items and biohazard trash (red color wrapped with double red color auto-cleavable bags)

should be placed adjacent to laminar flow only while working. During the non-working time, biohazard waste containers should be covered with la id and hidden near a corner.

5. Liquid CO_2 cylinders should be fastened with a chain adjacent to the incubator.
6. Incubator with shaker – used during enzymatic digestion of tissue
7. Sufficient racks and tubes should be present for placing sterile liquids.
8. The freezer and refrigerator (preferably glass window door) should be placed in one place.
9. A dedicated refrigerated centrifuge should be available in the cell culture lab.
10. An extra wash basin and collection tray should be present away from the laminar flow.
11. A dedicated trolley for carrying lab consumables should be present.

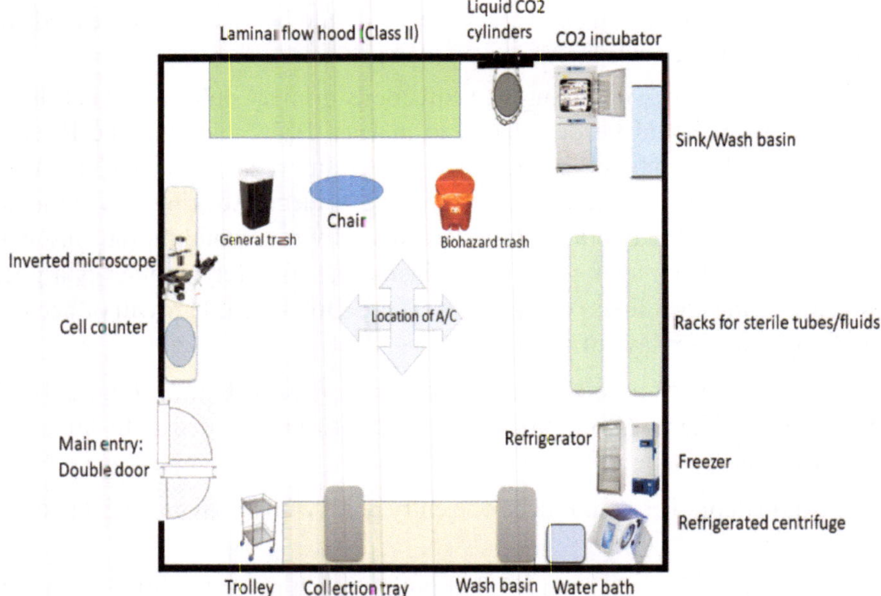

Figure 1: Layout of a cell culture lab.

Figure 2: An example of a mammalian cell lab showing arrangements of various equipment in the vicinity of the experimenter. (Photo credit: Ratan Choudhary's personal album from the University of Vermont, Burlington, USA).

Question

Q1. Draw a layout of the animal cell culture laboratory and explain the details about the location of different equipment present therein.

Ans. ....

....

....

....

....

....

....

....

....

....

....

....

....

....

Further readings

Culture of Animal Cells: A Manual of Basic Technique and Specialized Applications. By R. Ian Freshney. 2010. Print ISBN:9780470528129 |Online ISBN:9780470649367 |DOI:10.1002/9780470649367. 2010 John Wiley & Sons, Inc.

Observation and Notes

2

Basic Equipment of Animal Cell Culture Laboratories

Things required for cell culture

1. Lab – having negative pressure (HEPA filter)
2. Safety cabinet (Hood/laminar flow)
   a) Biosafety level 1 (BSL-1) – Basic level of protection. Agents handled under this cabinet are NOT infectious.
   b) BSL-2 – Appropriate for moderate risk agents. Most of the non-infectious cell cultures are done under this facility.

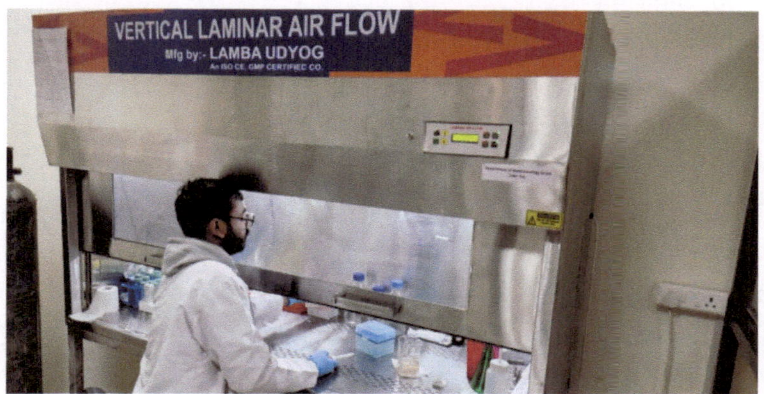

   c) BSL-3 – Potential infectious agents that can spread through aerosol and cause serious and lethal infections.
   d) BSL-4 – Exotic highly infectious agents that pose a high risk of life-threatening disease.
3. Inverted microscope Used to monitor growth of cells that decides confluency and splitting of cells during passage.

4. Refrigerated centrifuge

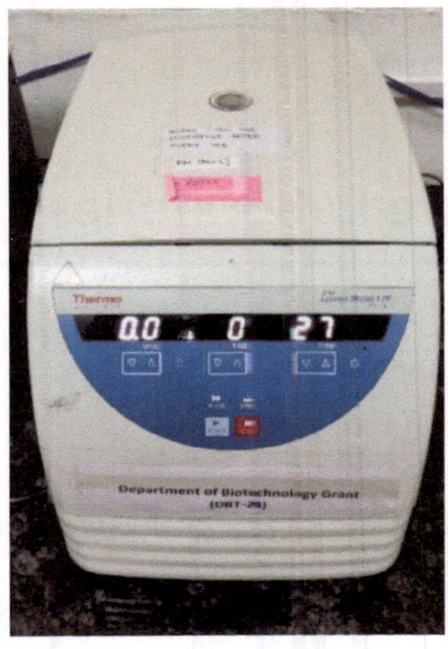

a) Table tops
b) Portable centrifuge
c) Big centrifuge
i) Non-refrigerated
ii) Refrigerated

5. Refrigerator – Glass door refrigerator well suited for keeping cell culture regents in refrigeration.
6. CO_2 Incubator – Following functions are done by the incubator,
   a) Incubator maintains the temperature at 37°C for the cell to grow
   b) 95% CO_2 and 5% O_2 level – CO_2 is required for buffering of culture medium
   c) It controls the humidity of the cells – 97% RH
   d) It maintains a sterile environment

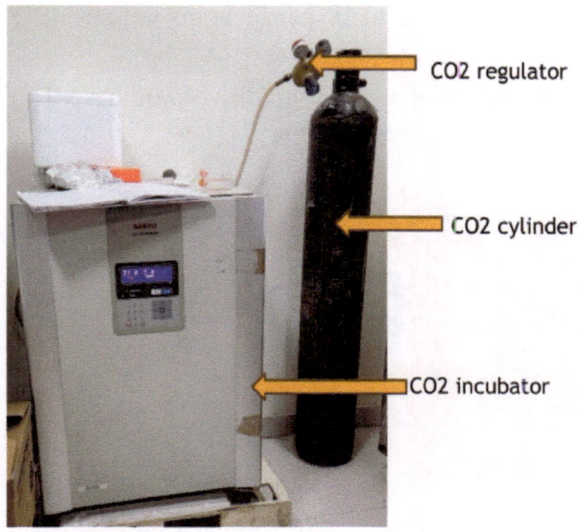

7. Water bath

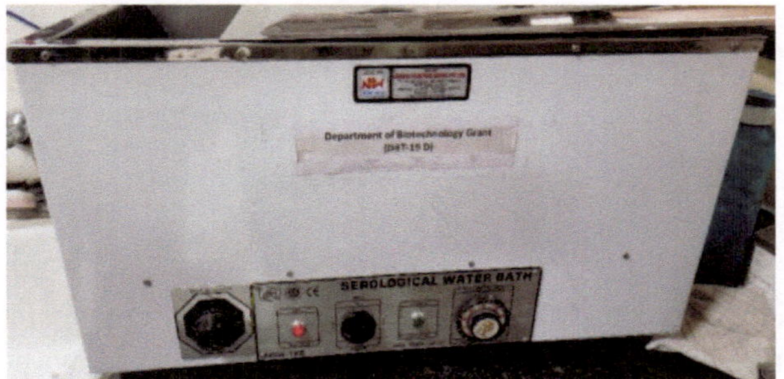

8. Autoclave – Required for sterilization of bottles, scissors, and buffers. We DONOT sterilize culture medium using autoclave.
9. Pipetman

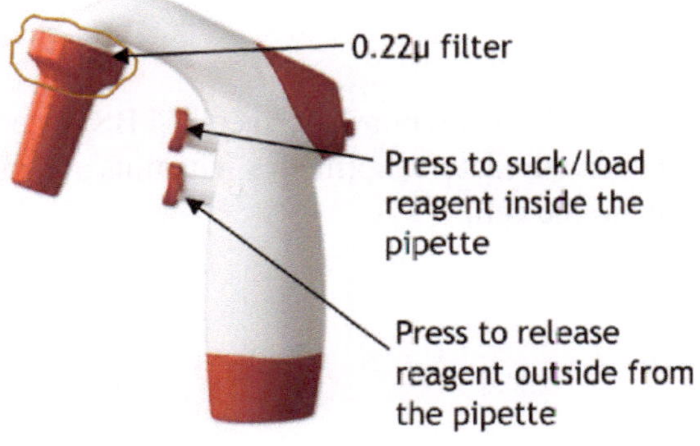

10. Glass/plastic pipettes - Used to dispense large volume (2-50 mL) of reagents.
11. Micropipettes - Used to dispense small volume (0.5-1000 μL) of reagents.

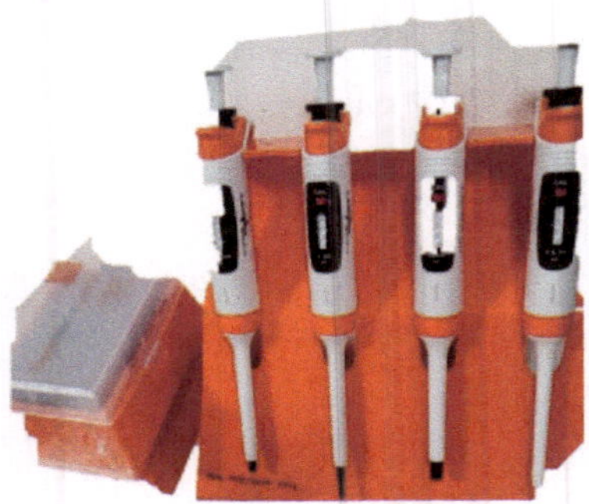

12. Cell culture medium - Medium to grow cells. Contains basal medium, fetal bovine serum, antibiotics and some other supplements depending on the cell types.

13. Washing buffer - Used to wash cells after before adding trypsin to cell during splitting. Also been used to wash tissue admixed with blood, making digestion medium and other purposes.

14. Fetal bovine serum (FBS) - Provides nutrients to cells. FBS is a complex fluid provides protein, growth factors, hormones, vitamins and minerals and attachment factors to the cell.

15. Apron -fully covered arms - It is a personal protective equipment.
16. Gloves - Working with sterile glove is MUST during cell culture.
17. Ethanol for wiping working space under the BSL.
18. Trypsin - It is a dissociation enzyme used to dislodge cells during splitting.
19. Cell counting (hemocytometer or cell counting machine) - Counting of cells are for deciding number of cells for seeding, cryopreservation and evaluating effect of treatment.

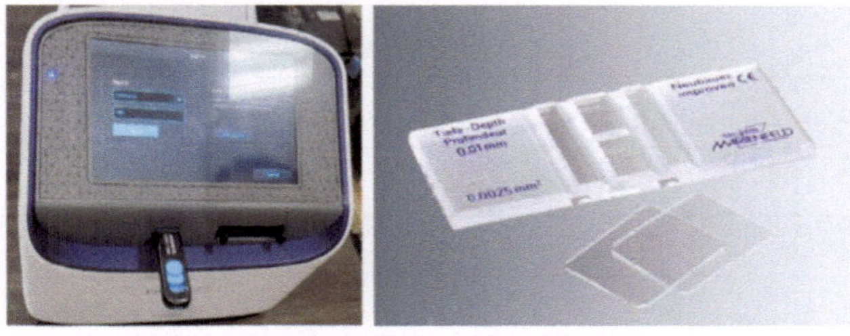

Automated cell counter (left) and hemocytometer (right)

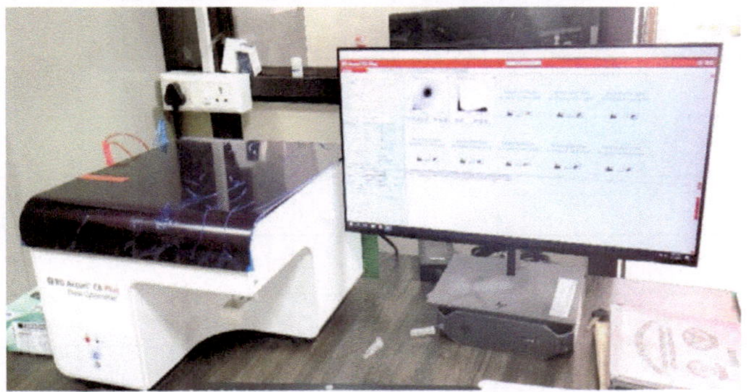

Figure: Flow cytometer can be the option equipment for more sophisticated analysis of cells.

Question

Q1. Write down the basic equipment required for cell culture work.

Ans. ....

....

Further readings

Culture of Animal Cells: A Manual of Basic Technique and Specialized Applications. By R. Ian Freshney. 2010. Print ISBN:9780470528129 |Online ISBN:9780470649367 |DOI:10.1002/9780470649367. 2010 John Wiley & Sons, Inc.

Paper on cell culture https://www.ncbi.nlm.nih.gov/pmc/articles/PMC7325[illegible]46/

Observation and Notes

3

Washing, Packing, and Sterilization of Glass and Plastic Wares

Working Principle: An autoclave is a sealed device that creates a different environment from the outside world under extraordinarily high pressure (Auto + clave = automatic locking). It uses moist heat or saturated steam under pressure to kill microorganisms and heat-resistant endospores. Its function is compared to that of a pressure cooker, i.e., steam is used to create a high-pressure environment. However temperature of autoclave or steam sterilizer is generally maintained at 121°C for 15-20 minutes that is achieved by raising the pressure inside the device. This high pressure increases the boiling point of water and therefore raises the temperature that is too high for the microbes to survive, so used for sterilization.

Types of Autoclave cycle

1. **Liquid cycle:** Liquids take time to sterilize as they have high heat capacity. It is based on the principle of pressurized steam at 15 psi, 121°C temperature for 20 minutes. It has latent heat, which is 7× heat 100°C. Liquids and biological waste require slow exhaust to prevent the boiling over of super-heated liquids.

   This excess temperature is enough to hydrolyze the proteins of bacteria, viruses, or any other contaminant and therefore kill it. However, wet heat is seven times more intense than dry heat, so it requires less temperature than the dry heat cycle of the autoclave.

2. **Dry Cycle:** Here, steam is the sterilization medium, which penetrates inside the load and makes contact with all the surfaces, so a fast exhaust cycle is used. It has a high temperature of 160°C for 60 minutes.

Here, protein hydrolysis does not take place. Instead, it kills all the microorganisms by oxidating their cellular contents. To kill the microorganism effectively, the heat required is higher.

3. **Pre vacuum Cycle:** It includes sterilization of the Pipette tip box; Biohazard waste should be in autoclave bags, Glassware that needs to be sterilized should be upright and Sharps decontaminated.

Things needed for autoclaving: 1. Personal protective equipment (PPE) that includes

- Lab coats
- Eye protection
- Face protection
- Closed toe shoes
- Heat resistant gloves to remove items especially hot glasswares

2. Dry cycle contents: Dry materials unwrapped or in porous wrap. It includes

- Metal
- Non-porous materials
- Empty glasswares

3. Wet cycle contents: Use glass containers with vented closures and 2/3 full only. It should include

- Aqueous solution
- Liquid media
- Non-flammable liquids
- Liquid biological waste

Steps of Autoclaving

Packaging and loading

1. Before using the autoclave, check inside for any items left by the previous user.
2. Place the items to be sterilized in the secondary container.
3. Do not overload or pack the bag too tightly. Leave sufficient room for steam circulation.
4. Place solid glassware inside the secondary container and autoclave in solid cycles.
5. For secondary containment, use autoclave trays made up of polypropylene, polycarbonate or stainless steel.
6. Choose the appropriate cycle for the material. Incorrect selection may cause damage to the autoclave.
7. A complete cycle takes 1 to 1.5 hours. Check chamber or jacket pressure gauge for min. Pressure of 20 pounds per square inch.
8. Close and lock the door.
9. Don't open the door while the autoclave is operating.

Precautions

1. Wear insulated thermal gloves when removing glassware from the autoclaves.

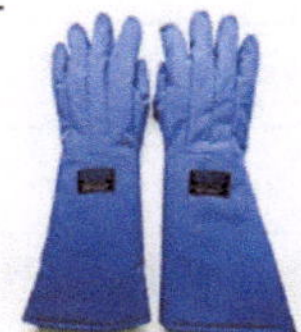

2. Ensure the employs using the autoclave should know how to use it safely.
3. All glassware needed to be inspected before autoclaving as older glassware is less stable and may break during the process.
4. Always wear PPE; gloves should be worn when handling autoclaved materials.

5. When sterilizing liquid, use the liquid cycle only.
6. Allow materials to cool before transporting.
7. Ensure all lids are loose and vented closures to prevent vacuum.
8. Sterilizers, racks, and materials would be very hot, so stand back from the sterilizer while opening the door.
9. Immediately clean any spilled materials.
10. Do not attempt to open the door when the cycle is in process.

Questions

Q1. Why do we sterilize things during the cell culture experiments?

Ans. ....

Q2. Write the principles of sterilization.

Ans. ....

Q3. Write down the various methods of autoclaving.

Ans. ....

Further reading

Culture of Animal Cells: A Manual of Basic Technique and Specialized Applications. By R. Ian Freshney. 2010. Print ISBN:9780470528129 |Online ISBN:9780470649367 |DOI:10.1002/9780470649367. 2010 John Wiley & Sons, Inc.

Observation and Notes

____

____

____

____

____

____

____

____

____

____

____

4

Preparation of Cell Culture Media and Reagents

In the realm of cell culture, the quality and composition of the growth media and reagents play a pivotal role in growth and sustaining cells in vitro. The preparation of these essential components requires precision, attention to detail, and adherence to the strict protocols to ensure optimal cellular growth, viability, and experimental reproducibility. This introduction serves as a foundational overview of the critical aspects involved in the preparation of media and reagents for cell culture experiments. Furthermore, the preparation of media and reagents for cell culture necessitates adherence to stringent quality control measures to ensure consistency and reproducibility across experiments. Validation of media batches, sterility testing, and periodic quality assessments are essential practices to mitigate variability and ensure the reliability of experimental results.

To Prepare Autoclavable Powdered Medium (1X)

1. In a container, add 950 mL of distilled water.
2. Add powdered medium to room temperature (15°C to 30°C) water with gentle stirring.
3. Cleanse the interior of the container to eliminate any remnants of powder.
4. Once fully dissolved, regulate the pH to a range of 4.1 to 4.2 using 1 N HCl.
5. Pour 95 mL of the solution into 100-mL bottles without sealing tightly.
6. Sterilize the medium in an autoclave for a minimum of 15 minutes at 121°C on a gradual exhaust cycle. However, considering variations in equipment performance, it is advisable to verify the autoclave sterilization process.
7. Under aseptic conditions, introduce 3 mL of 7.5% $NaHCO_3$ solution into each bottle and mix thoroughly.
8. Aseptically incorporate L-glutamine and antibiotic solutions, serum, or any other supplements as required. Ensure thorough mixing.

9. Aseptically adjust the pH of the medium to the desired final range of 7.2 to 7.4 by gradually adding 1 N NaOH or 1 N HCl while stirring if necessary.

To Prepare complete, supplemented tissue culture media

Serum and additional supplements are added to tissue culture media prior to use. FBS must be Heat inactivated before adding to media to eliminate complement activity. Make sure all reagents are in solution prior to addition to media. Media must be filtered through 0.22 um filter prior to use.

S.No.	Name	Description	Concentration	Storage
1	Cell media	MEM or DMEM	NA	4°C
2	Antibiotic solution	10000 U/mL Penicillin and 10000 U/mL Streptomycin	1% (final)	-20°C
3	HEPES	1M	1% (final)	4°C
4	L-glutamin	200 mM (100x)	1% (final)	4°C
5	Sodium pyruvate	10 mM (100x)	1% (final)	4°C
6	Fetal bovine serum (FBS)	Heat inactivated	5% or 10%	-80°C
7	Filter flasks	0.22 um Millipore	NA	Room temperature

Procedures

Obtain a 500 ml bottle of RPMI or DMEM media from walk-in 4°C cooler (26-38°C). Add half of the media to a 500 ml 0.22 um bottle top filter (with the vacuum turned off) Add the following:

5% FBS Media	10% FBS Media
a. 28 ml of FBS	a. 59 ml
b. 5.6 ml of pen-strep	b. 6.0 ml
c. 5.6 ml of L-Glutamine	c. 6.0 ml
d. 5.6 ml of HEPES buffer	d. 6.0 ml
e. 5.6 ml of sodium pyruvate	e. 6.0 ml

Apply vacuum to filter system and add the remaining media to the filter.

Question

Q1. What are the various components required to make a complete growth media?

Ans. ....

Further reading

Culture of Animal Cells: A Manual of Basic Technique and Specialized Applications. By R. Ian Freshney. 2010. Print ISBN:9780470528129 |Online ISBN:9780470649367 |DOI:10.1002/9780470649367. 2010 John Wiley & Sons, Inc.

5

Isolation of Primary Culture of Chicken Embryo Fibroblast

What is primary cell culture?

In vitro culture of cells freshly obtained from the tissue is called primary cell culture. Unlike that of cell line, primary cell culture grows for a limited period of time. Examples, are primary mammary epithelial cells, primary hepatocyte culture.

Aim: To prepare the primary cell culture from chicken embryos.

Materials required for the primary cell culture

- 70% alcohol
- Cotton swabs
- Small beaker 20-50 ml
- Forceps
- Straight and required curved,
- 9 cm Petri dishes
- 11 day embryonated eggs
- Balanced Salt solution (BSS)
- 250 ml conical flask
- Magnetic centrifuge tubes
- Haemocytometer
- Growth medium with serum
- Culture flasks

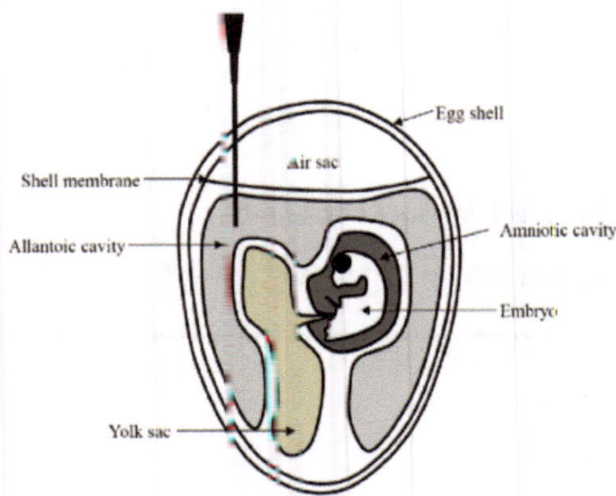

Source: 10.1007/978-1-4939-0758-8_22

Procedures

- Swab the embryonated eggs (11 days old) with 70% alcohol and place with blunt end uppermost in a small beaker.
- Break the top of the shell and peel off to the edge of the air sac with sterile forceps.
- Using a sterile forceps, remove the embryo from the shell and place it on a sterile Petri dish containing 20 ml BSS.
- Transfer to embryo to fresh sterile BSS and rinse
- Transfer to another dish, dissect off unwanted tissues and chop finely with scalpels to about 3mm diameter.
- Transfer by pipette to a 15-or 50 ml sterile centrifuge tube and allow the pieces to settle.
- Wash by resuspending the pieces in BSS, allowing the pieces to settle and removing the supernatant fluid, two or three times and then transfer the pieces into a 250 ml conical flask and add 100 ml of trypsin.
- Stir at about 200 rpm for 30 min at 37°C by using a magnetic stirrer.
- Allow pieces to settle collect supernatant, centrifuge at 1000 rpm for 5 min, resuspend pellet in 10 ml medium with serum and store cells on ice.
- Add fresh trypsin into the conical flask and continue to stir and incubate for a further 30 min Repeat steps 7 to 10 until no further disaggregation is apparent.
- Collect and pool chilled cell suspensions and count by hemocytometer.
- Dilute to 10^6 per ml on growth medium and seed as many flasks as are required with app.2 X 10^5 cells per cm^2.

- Change the medium at regular intervals 2 to 4 days as dictated by reduction in pH.
- Observe the flask under inverted microscope for the development of the cells.

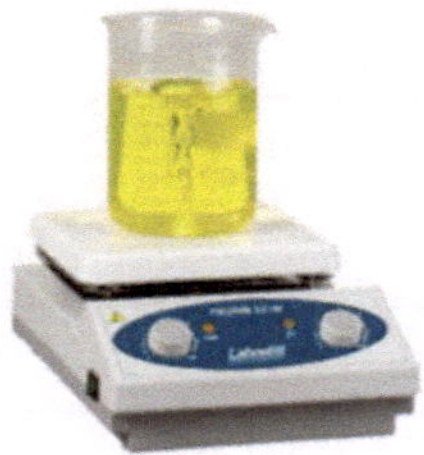

Questions

Q1. Write the morphology of the chicken embryo fibroblast observed under the microscope.

Ans. ....
....
....

Q2. Write down the procedures for the isolation of chicken embryo fibroblasts.

Ans. ....
....
....

Further readings

Culture of Animal Cells: A Manual of Basic Technique and Specialized Applications. By R. Ian Freshney. 2010. Print ISBN:9780470528129 |Online ISBN:9780470649367 |DOI:10.1002/9780470649367. 2010 John Wiley & Sons, Inc.

https://www.onlinebiologynotes.com/primary-cell-culture-preparation-of-primary-chick-embryo-fibroblast-cef-culture/ (Accessed on June 28 2021).

Observation and Notes

6

Isolation of Primary Culture of Canine Keratinocytes

Shipra Tiwari and Ratan Choudhary

What is keratinocyte?

The epidermal layer consists of keratinocytes, melanocytes, Langerhans cells, and Merkel cells. Their major function is to produce Keratinocytes which are the most abundant cells in the epidermis, keratin protein. Keratinocyte synthesizes major structural components of the epidermal barrier through a programmed process of differentiation. They are present in the innermost layer of the epidermis (Eckert *et al.*, 1989). The epidermis is composed of several layers, stratum corneum (horny layer), stratum lucidum (clear layer), stratum granulosum (granular layer), stratum spinosum (prickle cell layer), and stratum basale (basal layer), from the outermost to the deepest layer respectively. The process of maturation and movement of the keratinocytes from the stratum basale to the stratum corneum is what we call epidermal differentiation.

Aim: To prepare the primary cell culture of keratinocytes from canine skin

Table 1: Reagents used during the processing of canine skin tissue for the isolation of keratinocytes.

Components	Company	Catalog number	Pack size	Stock volume	Volume used
PBS	Sigma-Aldrich	D5652	50 L	1 L (1X)	50 ML
0.25%Trypsin-EDTA (1X)	Gibco	25200-072	500 ML	1 ML aliquots	1 ML
Complete medium: (DMEM + 10% FBS + 1% Ab)	HIMEDIA (DMEM)	AL183A	500 ML	50 ML complete medium	50 ML

Procedures for isolation of canine karatinocytes

When studying skin cells, isolating keratinocytes from tissue samples, such as skin biopsies, is a critical initial step. Here is a general guide on how to isolate keratinocytes:

1. Aseptically collect a skin tissue sample from a canine and transfer it into a 50 ml sterile tube containing PBS premixed with 2x concentration of antibiotics.
2. Keep the tube containing the skin sample on ice and transport it to the laboratory.
3. Shave and clean the skin sample with 70% ethanol and wash it thoroughly with PBS with 1x antibiotic solution to remove any extra blood stains and debris.
4. Prepare digestion media by mixing 1 ml of 0.25% trypsin + EDTA with 4 ml of PBS, resulting in 0.05% trypsin digestion media.
5. Place the skin tissue sample in a sterile petri plate and add 500-800 µl of digestion media. Begin chopping the tissue using scissors, blade, and forceps.
6. Once the tissue is chopped finely, transfer it to the remaining digestion media in the 50 ml tube.
7. Incubate the reaction in a water bath maintained at 37°C, with gentle shaking every 5-10 minutes.
8. After 30-45 minutes, neutralize the trypsin with complete media, pass the solution through a cell strainer, and centrifuge it at 2000 rpm for 5 minutes.
9. Resuspend the cell pellet in 1 ml of complete media, mix it with 4 ml of complete media, and seed it into a T25 flask for cultivation.
10. Monitor cell growth in a CO_2 incubator and initiate subculturing when optimal density is reached.
11. For trypsinization, aspirate the medium, rinse with PBS, add trypsin-EDTA, and incubate for 5-10 minutes.
12. Stop trypsin activity with complete media, collect, and centrifuge the cell suspension.
13. Resuspend the cell pellet in complete media for downstream applications.
14. Enrich keratinocytes by allowing them to adhere to a culture dish, wash to remove unattached cells, and detach adherent keratinocytes.
15. Centrifuge to obtain a purified keratinocyte pellet, resuspend it in a suitable medium for further analysis, or sub-culturing.
16. Perform Periodic Acid–Schiff Staining (PAS) to visualize glycogen and other structures in isolated keratinocytes.
17. Fix cells with buffered formalin, oxidize sugars with Periodic Acid, and develop color reaction with Modified Lillies Schiff Solution.

18. Stabilize staining with Sodium Borate, stain with acidified Lillie-Mayer hematoxylin, and dehydrate using graded alcohols.
19. Clear cells in xylene and mount them with synthetic resin for visualization and analysis.

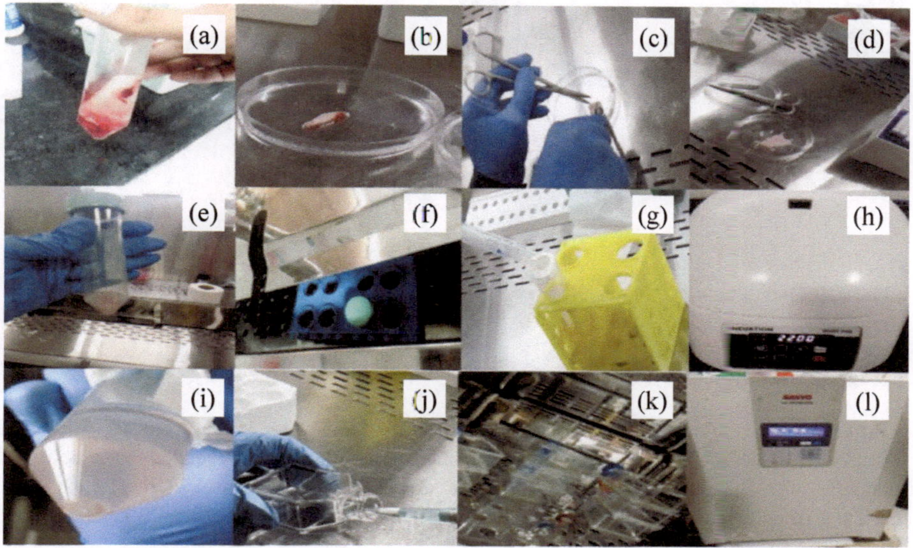

Figure 1: Stepwise process of isolating cells from canine skin tissue by trypsin digestion.

Top row (from left to right): **A)** Skin sample collected. **B)** Shaved, cleaned with ethanol, rinsed with PBS. **C)** Dissection of the tissue. **D)** Fine mincing of the tissue.

Middel row (from left to right): **E)** Addition of trypsin solution for digestion. **F)** Incubation in water bath at 37°C, 30 min. **G)** Filtration through 70 µm cell strainer. **H)** Centrifugation of the cell suspension (2200 rpm, 5min).

Bottom row (from left to right): **I)** Mixed cell pellet obtained after centrifugation. **J)** Seeding of the cells in T25 culture flask with complete media. **K)** Incubation of the cells at 37°C in 5% CO_2 incubator. **L)** Culture in 5% CO incubator.

Cell enrichment through a differential plating technique was successfully implemented, demonstrating the preferential adhesion of keratinocytes to a coated culture dish. This method effectively separated the keratinocytes from other cell types, such as fibroblasts, enhancing the purity of the cell population.

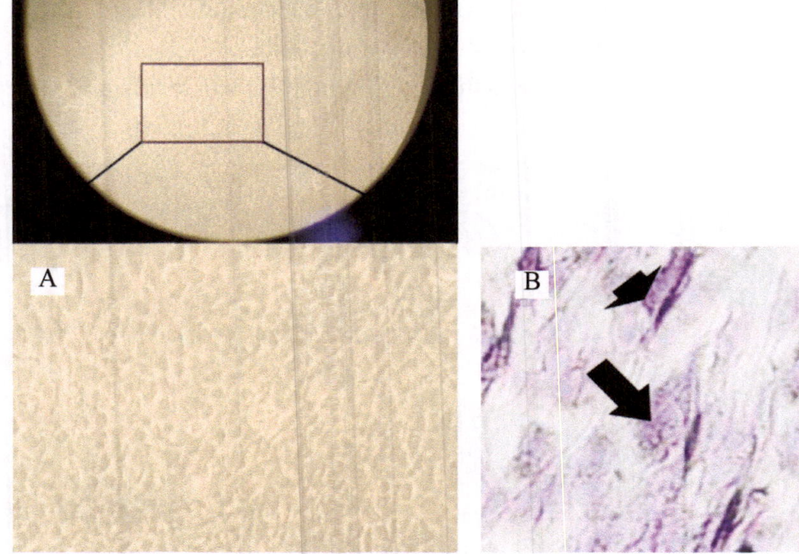

Figure 2: Morphology of canine keratinocytes in culture. **A)** Mixed cells of skin were differentially seeded, and after 2 hours of seeding, supernatants were reseeded into a new well plate. Because of the differential adhesion of keratinocytes and fibroblasts, both cell types were enriched in culture. **B)** PAS-stained long and elongated fibroblast (arrowhead) cells and oval shaped keratinocytes (arrow) can be seen.

Questions

Q1. What are the key considerations and challenges in maintaining the purity and viability of isolated keratinocyte populations during the isolation process?

Ans. ....

....

Q2. What is the rationale behind using 70% ethanol to clean the skin sample before proceeding with the isolation protocol?

Ans. ....

....

Further readings

Serra, M., Brazís, P., Puigdemont, A., Fondevila, D., Romano, V., Torre, C., & Ferrer, L. (2007). Development and characterization of a canine skin equivalent. Experimental dermatology, 16(2), 135–142.

Vieira, N. M., Brandalise, V., Zucconi, E., Secco, M., Strauss, B. E., & Zatz, M. (2010). Isolation, characterization, and differentiation potential of canine adipose-derived stem cells. Cell transplantation, 19(3), 279–289.

7

Isolation of Primary Culture of Canine Fibroblast

Samreet Kaur and Ratan Choudhary

What is dermal fibroblast?

Skin fibroblasts are a type of cell found in the dermis, the middle layer of the skin. They are the most common cells in connective tissue and play a crucial role in maintaining the structural integrity of the skin. Fibroblasts are responsible for synthesizing and secreting extracellular matrix proteins such as collagen, elastin, and fibronectin, which provide support and elasticity to the skin. Additionally, fibroblasts play a role in wound healing by proliferating and migrating to the site of injury, where they aid in tissue repair and scar formation.

Aim: To prepare the primary cell culture of fibroblasts from canine skin.

Tools and instrument required for the procedure

1. Tissue collection box with ice cubes - To prevent contamination and maintain the stable temperature
2. Surgical Gloves - To prevent contamination
3. Tissue collection tube with 1X PBS - To collect sample
4. Laminar flow hood - To work in a sterile environment and prevent contamination
5. Cell culture dish - To digest the skin
6. Sterile 1000 µL and 200 µL pipette and tips-For accurate measurement and transfer of cell suspension and culture media.
7. Forceps - Fine-tipped forceps for handling and manipulating tissue.
8. Scalpel and blade holder - Scalpel for the dissection of skin tissue.
9. Surgical scissors - Chopping of the skin tissue
10. Falcon 70 µm cell strainer - To filter out undigested tissue debris and obtain a homogeneous cell suspension.
11. 15 ml and 50 ml Centrifuge tubes - To collect and centrifuge the digested cell suspension.

12. Water bath (maintained at 37°C) - To maintain the enzymatic solution at the desired temperature during the digestion process.
13. Centrifuge For cell pellet formation and separation of cells from the digested tissue.
14. CO_2 incubator - To provide a controlled environment for cell culture, including temperature and CO_2 levels.
15. T25 flask - To culture the isolated cells
16. Serological pipette - Used to transfer culture medium
17. Waste container - A container for the disposal of used pipette tips and other disposable items.

Regents required for procedure

1. 70% ethanol - To prevent contamination
2. 1x Phosphate buffer saline (PBS) - Used for rinsing and washing tissues and cells during the digestion process.
3. Dulbecco's Modified Eagle Medium (DMEM) - A medium that provides nutrients and a suitable environment for cell survival during the digestion process.
4. Enzyme solution - Collagenase type I 0.1 mg/ml in PBS An enzyme that breaks down collagen.
5. Culture medium DMEM + 10% FBS– 1% AB - To support the growth, maintenance, and propagation of microorganisms, cells, or tissues.
6. Fetal Bovine serum (FBS) - To neutralize enzymes and provide additional nutrients for cell survival.
7. Antibiotics (AB) - Penicillin and streptomycin are commonly added to the medium to prevent bacterial and fungal contamination

Procedures of isolation of canine fibroblasts are similar to the procedures for isolation of canine karatinocytes dicussed in earlier chapter.

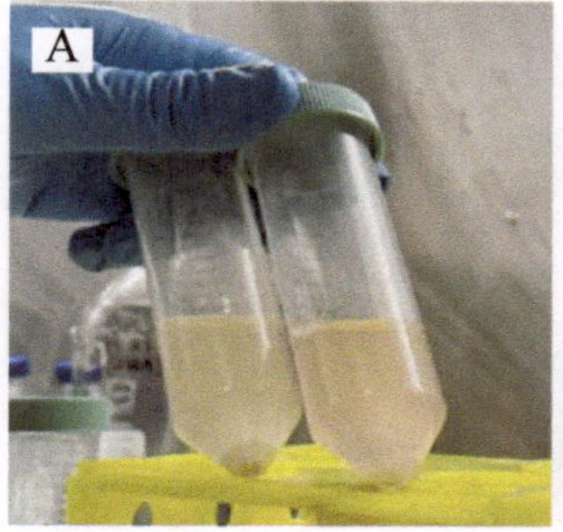

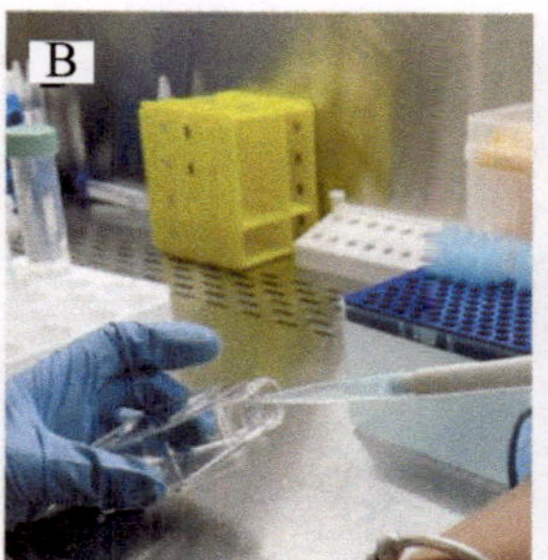

Figure: Isolation of canine fibroblast cells from canine skin. **A)** Mixed cell pellet obtained after centrifugation. **B)** Seeding of mixed cells in T25 culture flask with complete media. **C)** Incubation of the cells at 37°C in 5% CO_2 incubator

Cell enrichment through a differential plating technique was successfully implemented, demonstrating the preferential adhesion of keratinocytes to a coated culture dish. This method effectively separated the keratinocytes from other cell types, such as fibroblasts, enhancing the purity of the cell population. Leveraging the inherent tendency of fibroblasts to adhere to the culture surface, we employed a serial differential plating technique. Every two hours, unattached cells, primarily keratinocytes due to their suspension-favoring morphology, were transferred to a new well of a 6-well culture plate. After six rounds of this selective transfer, the majority of fibroblasts remained attached, depleting their population with each transfer. Conversely, the keratinocyte population progressively enriched, resulting in a culture enriched for these desired cells.

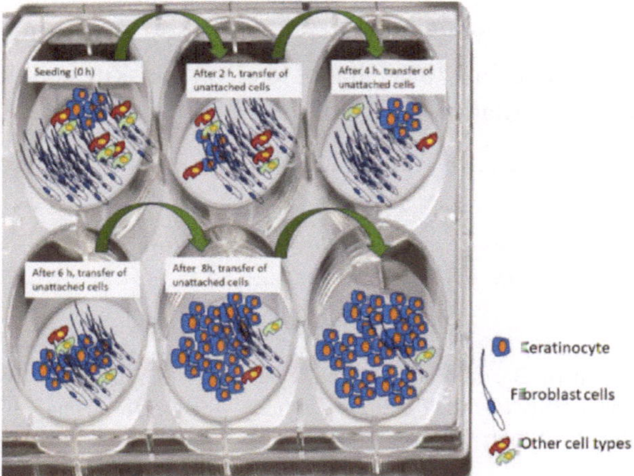

Figure: Cell enrichment through a differential plating technique. Through a separation process, the population of keratinocytes progressively increased, while the fibroblast population decreased. The initial well yielded a culture enriched with fibroblasts, while the final well provided a harvest of primarily keratinocytes.

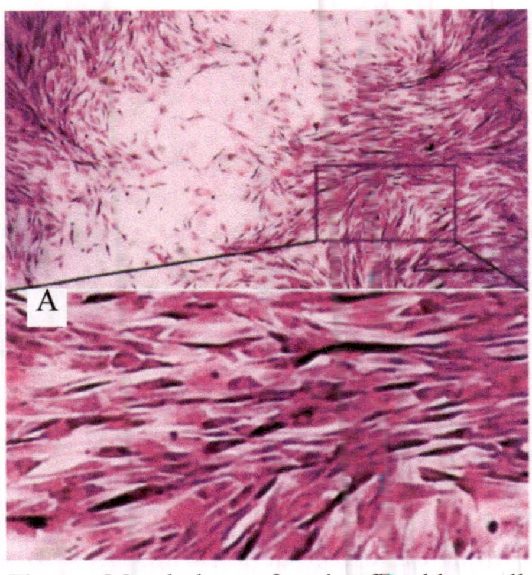

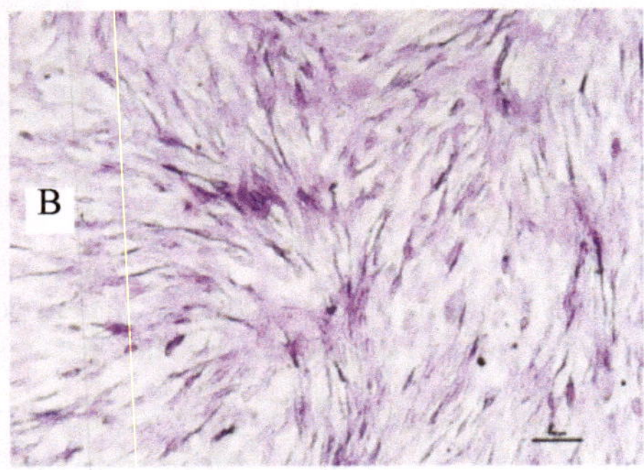

Figure: Morphology of canine fibroblast cells in culture. **A)** Hematoxyline staining of canine fibroblast after one week of seeding, **B)** PAS-stained long and elongated fibroblast cells can be seen.

Questions

Q1. What are the key considerations and challenges in maintaining the purity and viability of isolated keratinocyte populations during the isolation process?

Ans. ....

....

....

....

Q2. What is the rationale behind using 70% ethanol to clean the skin sample before proceeding with the isolation protocol?

Ans. ....

....

....

....

Further readings

Culture of Animal Cells: A Manual of Basic Technique and Specialized Applications. By R. Ian Freshney. 2010. Print ISBN:9780470528129 |Online ISBN:9780470649367 |DOI:10.1002/9780470649367. 2010 John Wiley & Sons, Inc.

8

Culture and Sub-culturing of Continuous Cell Lines

What is subculture?

In cell culture, a process called subculturing, or passaging, keeps cells alive and growing. It involves taking cells from an existing container, removing the used medium, and placing them in a new container with fresh nutrients. This allows the cells to continue multiplying and creates more cells for research.

Adherent subculture protocol

Cells nearing 80% coverage need splitting!

1. Warm things up: Get your media, detachment solution, and incubator ready at 98.6°F (37°C).
2. Safety first! Disinfect your work area and supplies inside a biosafety cabinet.
3. Gently rinse: Remove the old media and wash the cells with a salt solution at room temperature. Be careful not to dislodge the cells.
4. Detach the cells: Add a pre-warmed enzyme solution to detach the cells from the flask surface. Keep an eye on them to make sure it works (usually 2 minutes). Note: Not all cells need this enzyme, and it can be harmful to some.
5. Stop the enzyme: Once all the cells are detached, add a culture medium with serum to neutralize the enzyme.
6. Collect the cells: Transfer the cell suspension to a tube, rinse the flask with fresh medium to get all the cells, and add it to the same tube.
7. Spin it down: Centrifuge the cell suspension for 5 minutes at room temperature (1000 rpm).
8. Get ready for counting: Remove the liquid, loosen the cell clump, and resuspend the cells in a fresh medium at a good volume for counting.
9. Count your cells: Use a hemocytometer to count the cells and determine their health.

10. Seeding for growth: Based on your cell count, dilute the cell suspension and add it to a new flask at the right density. Don't forget to label the flask with cell line, passage number, and other important info.
11. Back to the incubator: Place your newly seeded flask in a warm, humidified incubator with 5% CO_2.

Sub-culturing loosely attached cell lines requiring cell scraping for sub-culture

1. Remove Used Media Safely: First, carefully dispose of the old growth medium from your cell flask. Pour it slowly into a waste container with diluted bleach (around 10% sodium hypochlorite) to avoid spills and contamination.
2. Refresh the Cells: Right after removing the old medium, gently add an equal amount of fresh, pre-warmed culture medium to the flask.
3. Transfer Cells for Seeding: Use a sterile pipette to take out a specific amount of cell suspension based on your desired split ratio. Here's a guide:
4. For a 1:2 split (meaning you'll dilute the cells 1 part cells to 2 parts new media), take 50 ml of cell suspension from a 100 ml flask and transfer it to a new flask.
5. Adjust the volume you take out for different split ratios (e.g., 20 ml for a 1:5 split, 10 ml for 1:10 split from a 100 ml flask).
6. Fill New Flasks with Fresh Media: Depending on the size of your new flask, add enough pre-warmed culture medium to reach the recommended final volume. Here's a general guideline:

Small flasks (25 cm²): 5-10 ml

Medium flasks (75 cm²): 10-30 ml

Large flasks (175 cm²): 40-150 ml

By following these steps, you'll have successfully changed the media, transferred cells for propagation, and prepared them for further growth in new flasks.

Questions

Q1. What is the role of trypsin in subculture?

Ans. ....
....
....
....
....
....
....
....

Q2. How do we neutralize action of trypsin? Is there any other alternative of trypsin for the dissociation of cells?

Ans. ....
....
....
....
....
....
....
....

Further readings

Culture of Animal Cells: A Manual of Basic Technique and Specialized Applications. By R. Ian Freshney. 2010. Print ISBN:9780470528129 |Online ISBN:9780470649367 |DOI:10.1002/9780470649367. 2010 John Wiley & Sons, Inc.

https://www.abcam.com/protocols/mammalian-cell-tissue-culture-techniques-protocol (accessed on June 28 2021).

Observation and Notes

9

Trypan Blue Dye Exclusion Cell Viability Assay

Cell counting is the fundamental techniques in various fields of biology, medicine and biotechnology.

Importance of cell counting

1. Research: Cell counting is a critical component of biological and medical research. It helps scientists understand the behavior of cells in various contexts, such as studying cell proliferation, differentiation, and response to stimuli.
2. Disease diagnosis: Monitoring various medical conditions, including infections, autoimmune diseases, cancer and blood disorders. Complete blood counts (CBCs) are routinely used to assess the levels of different blood cell types and assess disease conditions.
3. Stem Cell Research: Cell counting plays a significant role in stem cell research, where precise quantification of stem cells is necessary for various applications, including regenerative medicine and tissue engineering.
4. Cell Culture: Cell counting is integral part to determine seeding density. Seeding of cells to available surface area for growth is important consideration for successful culture.
5. Cell-Based Assay: The cell-based assays are used for drug screening, toxicity testing, and evaluating the efficacy of new compounds. Accurate cell counts are essential to ensure the reliability of assay results.

Methods of cell counting

- *Manual counting*

  a) Hemocytometer

- *Automated counting*

  a) Cell counters - These are instruments designed to count cells automatically, saving time and reducing user variability. Example, Countess 3 Automated Cell Counter from Invitrogen.

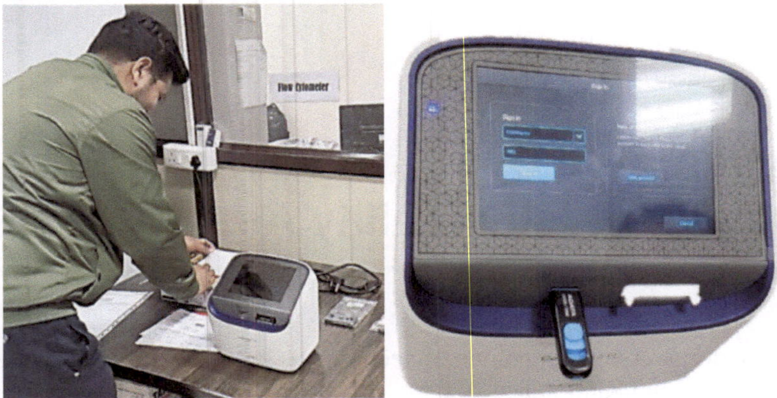

Figure: Man installing the automated cell counter in the lab. Right panel shows an automated cell counter

b) Flow cytometer - Flow cytometry is a powerful technique that allow cell counting as well as characterizes various parameters of cells like size, shaper, roundness and others.

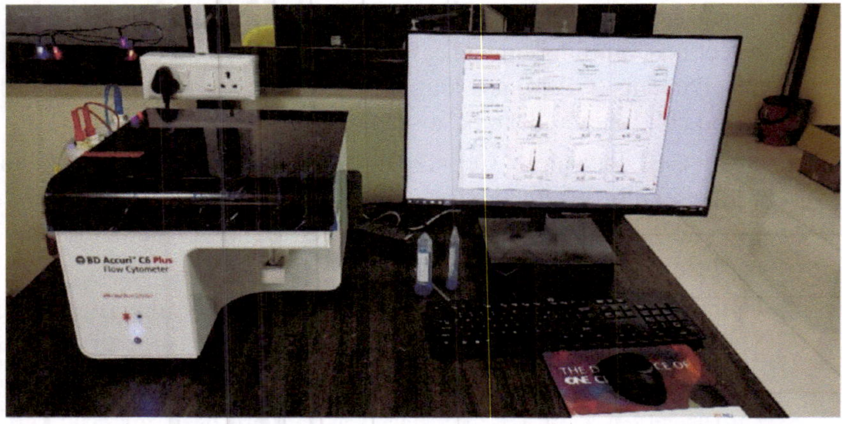

Figure: An example of flow cytometer (BD C6 Accuri) for automated counting (photo credit: Animal Stem Cells Lab, College of Animal Biotechnology, GADVASU, Ludhiana)

c) FACS – Advanced flow cytometer that allows isolation and culture of specific cell population from the mixture of cells.

d) Image analysis software: In combination with a standard light microscope, image analysis software can be used to count cells manually from captured images. This method provides a compromise between manual and automated counting and is suitable for various cell types.

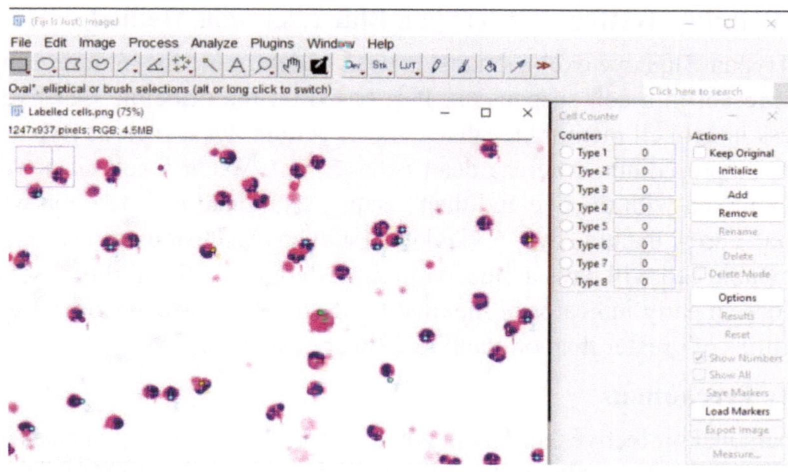

Figure: Counting different cell types using image analysis software (Image). It allow manual and automated counting of cells of captured photomicrographs of cells, cell smears and tissue sections.

Stains of cell counting

Stains are used in cell counting and microscopy to improve the visibility of cells, differentiate between live and dead cells, or highlight specific cell structures or components. Here are some common types of stains used in cell counting:

1. Viability stain
   a) Trypan Blue: Trypan blue is a widely used viability stain that can be excluded by live cells with intact cell membranes. Dead cells, with compromised membranes, take up the stain and appear blue when viewed under a microscope.
   b) Propidium Iodide (PI): PI is a red fluorescent dye that penetrates the membranes of dead or damaged cells. Live cells with intact membranes exclude PI. It is often used in flow cytometry and fluorescence microscopy.
2. Nuclear stain
   a) DAPI (4',6-diamidino-2-phenylindole): DAPI is a blue fluorescent dye, binds to DNA in the cell nucleus. It is used to count cells based on the number of stained nuclei.
   b) Hoechst Stains: Hoechst (pronounce as hoh-chayst) dyes are similar to DAPI and are often used to stain nuclei for cell counting.
3. Cytoplasmic stain

Cell Viability Testing with Trypan Blue Exclusion Method

The Trypan Blue dye exclusion test is used to determine the number of viable cells present in a cell suspension. It is based on the principle that live cells possess intact cell membranes that exclude certain dyes, such as trypan blue, Eosin, or propidium, whereas dead cells do not. When a cell suspension is simply mixed with the dye and then visually examined to determine whether cells take up or exclude dye. A viable cell will have a clear cytoplasm whereas a nonviable cell will have a blue cytoplasm. Periodic cell viability assessment provides an early indicator of the quality of your fresh cells prior to freezing. Viabilities of greater than or equal to 95% are excellent.

Safety Precautions

Use personal protective equipment when performing this assay, such as gloves and a lab coat. According to the Material Safety Data Sheet (MSDS), trypan blue may cause cancer, so practice appropriate laboratory safety methods.

Equipment and Supplies

- Pipette and tips
- Trypan Blue
- Hemocytometer and coverslip
- Cryovials
- Microscope Counter

Figure: Microscope counter. Each time you see cells under microscope, press the key to enter it to the total counting.

Procedure of counting cells

1. Before the cells have a chance to settle, take out 0.5 mL of cell suspension using a 5 mL sterile pipette and place in an Eppendorf tube.
2. Take 100 µL of cells into a new Eppendorf tube and add 400 µL 0.4% Trypan Blue (final concentration 0.32%; 1 part cell suspension + 4 part dye thus, 5 is the dilution factor). Mix gently.
3. Incubate mixture for less than three minutes at room temperature. If cells are counted after approximately five minutes, viability will be inaccurate due to cell death

4. With the cover slip already in place, fill one side of a hemacytometer counter with the cell suspension by placing the tip of the pipette at the notch. Typically, each side will take 10 to 20 µl.

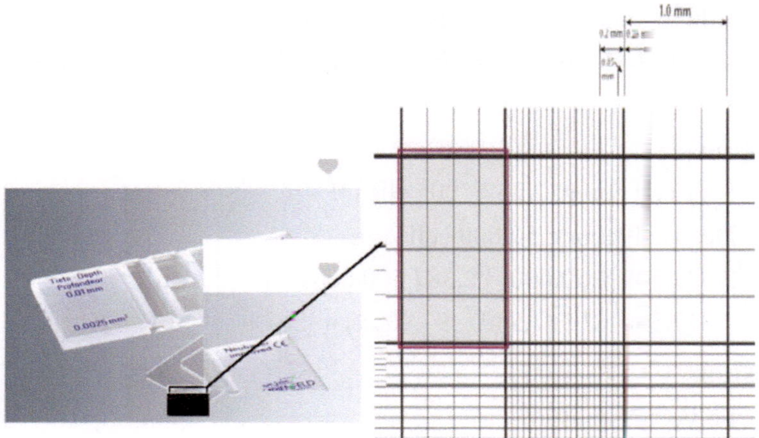

5. Place the hemacytometer on the stage of a light microscope and focus onto the cells.
6. To calculate the number of viable cells/mL:

Take the average cell count from each of the sets of 16 corner squares.

Multiply by 10,000 (104).

Multiply by 5 to correct for the 1:5 dilution from the Trypan Blue addition. If cells are diluted with trypan blue in 1:1 ratio, then multiply by 2.

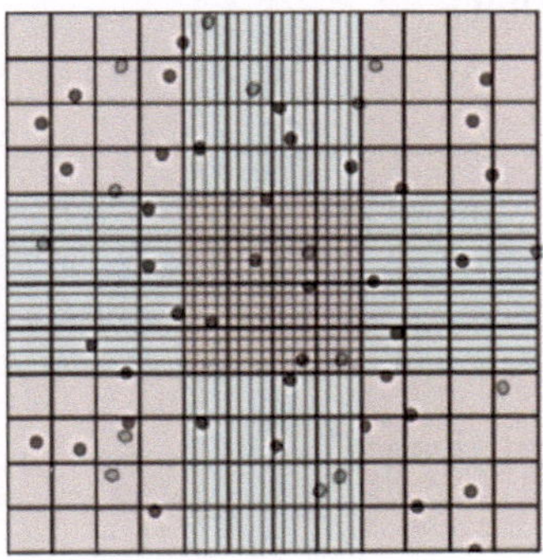

Example: If the cell counts for each of the 16 squares were 50, 40, 45, 52, the average cell count would be:

(50 + 40 + 45 +52) ÷ 4 = 46.75

46.75 x 10,000 (104) = 467,500

467,500 x 5 = 2,337,500 live cells/mL in original cell suspension

To calculate viability

If both live and dead cell counts have been recorded for each set of 16 corner squares, an estimate viability can be calculated.

1. Add together the live and dead cell count to obtain a total cell count.
2. Divide the live cell count by the total cell count to calculate the percentage viability.

Example

- Live cell count: 2,337,500 cells/mL
- Dead cell count: 50,000 cells/mL
- 2,337,500 + 50,000 = 2,387,500 cells
- 2,337,500 ÷2,387,500 = 97.9% viability

Question

Q1. Solve this question

Total no cells present in four 16 squares are 20, 22, 25 and 24 and the dilution factor is 4 then

a) What is the total no of cells present in 5 ml total cell suspension vial?

b) Of these, there are total 20000 cells are dead (appears blue in all four 16 squares) then calculate the percentage viability of the cells.

Ans. ....

....

Q2. Define principles of live and deal cell counting.

Ans. ....

....

Q3. What are the various types of stains used in cell counting?

Ans. ....

....

Q4. Enlist methods of cell counting and its significance.

Ans. ....

....

Further readings

Strober W. Trypan Blue Exclusion Test of Cell Viability. Curr Protoc Immunol. 2015 Nov 2;111:A3.B.1-A3.B.3. doi: 10.1002/0471142735.ima03bs111.

https://www.sigmaaldrich.com/deepweb/assets/sigmaaldrich/marketing/global/documents/331/916/t8154use.pdf (Accessed on June 28 2021).

10

Isolation and Culture Techniques for Lymphocytes

Paramjeet Sharma and Ratan Choudhary

Introduction

Peripheral blood mononuclear cells (PBMCs) consist of chiefly of lymphocytes and monocytes. Purified PBMCs are used in vitro to evaluate a variety of functions of lymphocytes viz; a) proliferation to mitogenic (PHA, Con-A) stimulation, b) monitoring of prior sensitization in antigen recall assays by scoring lymphocyte proliferation, c) immunophenotyping for surface markers as well as intracellular molecules in monocytes and lymphocytes etc. Activation of monocytes/macrophages by small molecules, cytokines and pathogen components can also be monitored. PBMCs can also be used for a variety of structural and functional studies for addressing issues in human immunology such as scoring for apoptosis and production of cytokines as well as other mediators in vitro.

Background information

Blood is a liquid mixture comprised of plasma, cells and proteins. The liquid portion, plasma, makes up half of the blood volume, and cells the other half. Isolation of lymphocytes is a common procedure in immunology and cell biology that involves separating and collecting lymphocytes.

Types of blood cells include

1. Red blood cells (RBCs) - carry oxygen to tissues
2. Platelets that help blood clot
3. Peripheral blood mononuclear cells (PBMCs) and Granulocytes, which help fight infection.
   a) PBMCs are a mixed population of myeloid and lymphoid cells. The below figure illustrates the cellular composition of whole blood as well as the various myeloid and lymphoid cells.
   b) Monocytes generally are end cells and do not proliferate. In absence of mitogens the proliferation of PBMCs will be negligible.

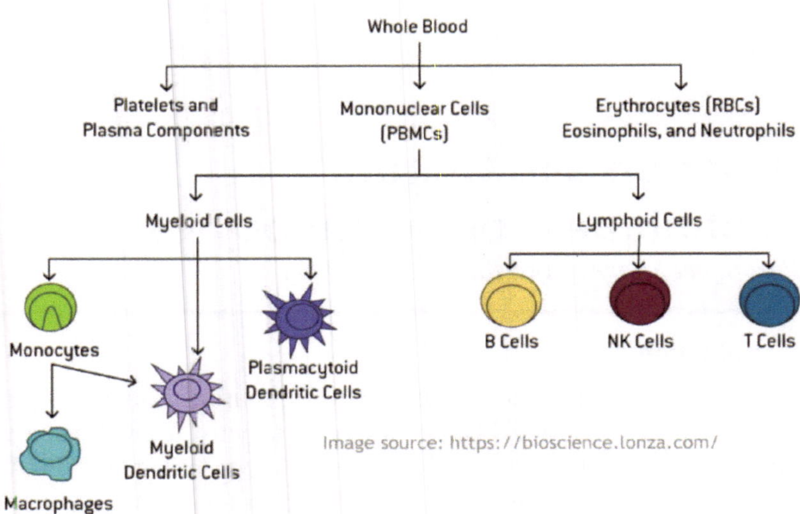

Significance of Lymphocyte Isolation

There are several reasons why isolating lymphocytes is significant:

Immunological Research: Isolating lymphocytes is essential for studying their functions and roles in the immune system. Researchers can use isolated lymphocytes to investigate how they respond to pathogens.

Diagnostic Testing: Isolated lymphocytes can be used in diagnostic tests to assess immune status. For example, tumor-infiltrating T-lymphocyte (TIL) subsets in canine oral melanoma is relevant to predict tumor aggressiveness and patient prognosis (Yasumaru *et al.*, 2021).

Cell Therapy: In the field of cell therapy, particularly for cancer treatment, lymphocytes can be isolated and manipulated for adoptive immunotherapy. Techniques like chimeric antigen receptor (CAR) T-cell (or T immune cells) therapy involve isolating and genetically modifying a patient's own T lymphocytes to enhance their ability to target and destroy cancer cells. In human, various FDA has approved multiple human CAR T therapies, however in veterinary oncology, promise of the therapy remain elusive.

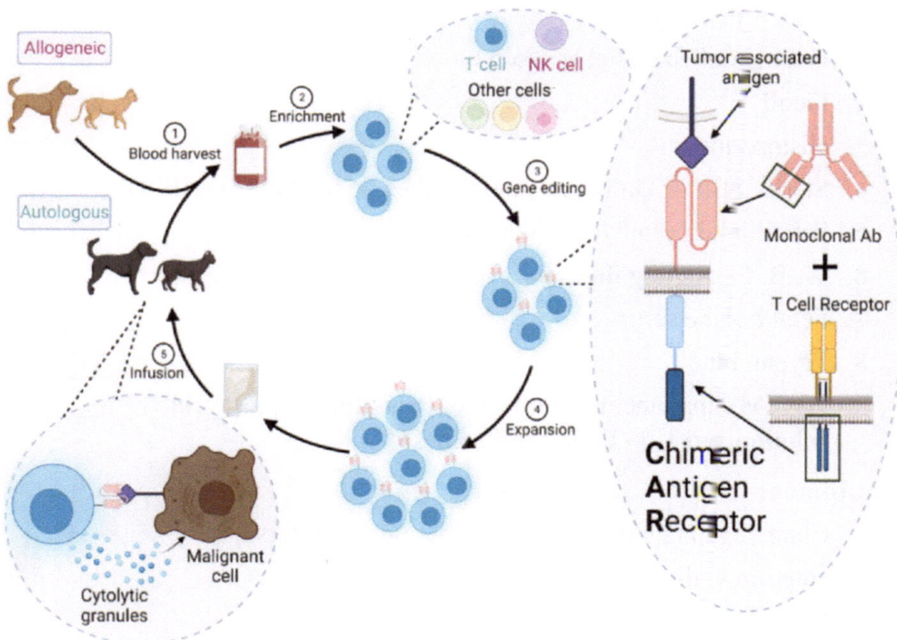

Figure: A schematic diagram showing CAR-T therapy in veterinary medicine. T-cells harvested from blood, is engineered to express CAR. The CAR contains variable heavy and light chain of monoclonal antibodies specific to tumor antigen. CAR-T cells, when injected in to the patient, detect and destroy tumor cells expressing tumor antigens. (Adapted from Cockey and Leifer, 2023).

Transplantation: Lymphocyte isolation is crucial in assessing the compatibility of donors and recipients in organ and bone marrow transplantation in humans.

Drug Development: The isolation of lymphocytes is significant in drug development and testing. Isolated lymphocytes are used to screen potential drugs or therapies (including stem cell therapy) for their effects on immune responses and cell function. Co-culture of mesenchymal stem cells with lymphocytes is been use to evaluate immunomodulatory potential of stem cells (Hitesh Rana, MSc thesis 2023).

Common methods for isolating lymphocytes include gradient centrifugation, immunomagnetic separation, and fluorescence-activated cell sorting (FACS). Each method has its advantages and is chosen based on the specific research or clinical application. In this chapter, we will focus on isolation of canine lymphocytes using density gradient centrifugation.

Materials and Reagents

1. Freshly collected heparinised blood
2. Ficoll
3. Histopaque
4. Sterile PBS or Dulbecco's modified eagle medium (DMEM)
5. Pencillin-streptomycin solution
6. W. B. C. diluting fluid
7. Fetal bovine serum
8. Trypan blue
9. DMEM supplemented with 1% of Pencillin-streptomycin solution and 10% FBS (see Recipes)

Equipment

1. Centrifuge machine with swing-out bucket rotors
2. Heparin vials
3. Sterile 15 ml centrifuge tube
4. Auto pipettes
5. 200 µl and 1 ml tips
6. 24 well cell culture plate
7. Haemocytometer
8. Tissue culture hood
9. CO_2 incubator
10. Microscope

Procedure

1. Collect about 5-10 ml of blood sample (examples, tail vein from cow, saphenous vein from dog, Jugular vein from goat) in heparinised vials and mix well by gently inverting the tubes several times.
2. Isolate bovine PBMCs by gradient centrifugation using Ficoll-Histophaque.
3. Wash cells (centrifuge at 100 x g for 10 min) with 10 ml of sterile DMEM (without FBS) twice.

Note: Cold DMEM is not used routinely for washing lymphocytes from culture cavities while setting up cultures. Rather when monocytes bound tightly to plastic cavities are needed to be harvested pre-chilled DMEM can be used.

4. Discard medium and re-suspend the cell pellet in 1 ml of sterile Dulbecco's modified eagle medium.

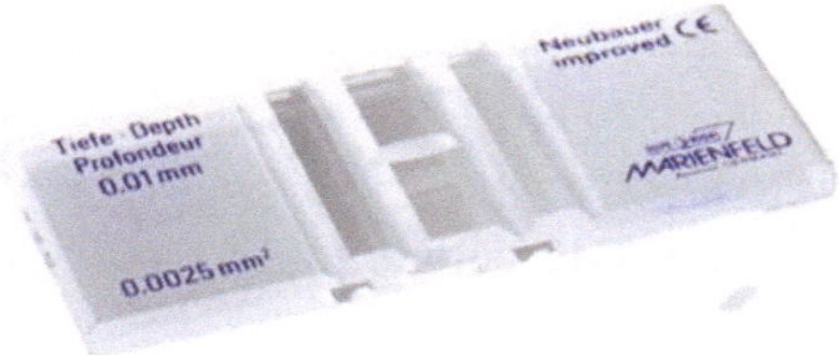

5. Count cells by haemocytometer using W.B.C. diluting fluid: Add 10 µl of cell suspension to 190 µl of W.B.C. diluting fluid and mix well (see practical no. 6: Viability assay by trypan blue dye exclusion method" for counting of cells using hemocytometer)

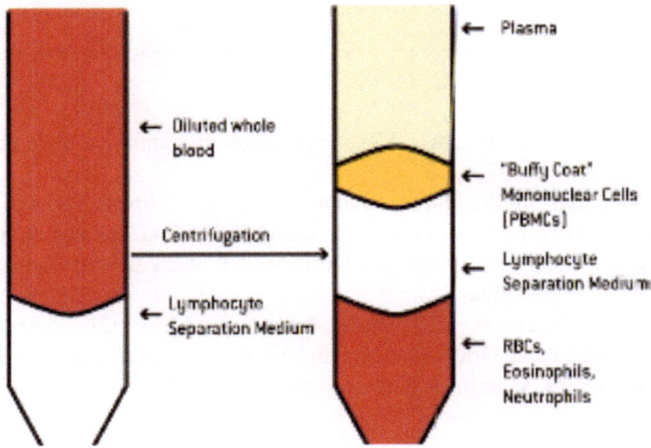

6. Seed 500 µl of cell suspension in a 24 well culture plate.

   Note: Monocytes in PBMCs get attached to the plastic in about 2-3 h when incubated at 37°C.

7. Cells can be treated with different antigens for different period of times and the supernatants can be analysed for cytokine levels. The cells can be analysed for phenotypic change, apoptosis or proliferation.

   Note: PBMCs are primary cells and cannot be cultured for more than one passage under normal conditions.

Questions

Q1. What are the various methods of isolation of PBMC? Describe any one of the methodologies.

Ans. ....

....

....

Q2. Write down the applications of culturing PBMC.

Ans. ....

....

....

Q3. Why mitogens are added into the PBMC culture?

Ans. ....

....

....

Q4. Draw morphology of various types of PBMC.

Ans. ....

....

....

Further reading

Cockey, J. R., & Leifer, C. A. (2023). Racing CARs to veterinary immuno-oncology. Frontiers in Veterinary Science, 10, 1130182.

Hitesh Rana. (2023). Evaluation of Immunomodulatory Effects of Adipose Tissue Derived Stromal Vascular Fraction (AT-SVFs) on Canine Peripheral Blood Mononuclear Cells. MSc thesis. College of Animal Biotechnology, Guru Angad Dev Veterinary and Animal Sciences University, Ludhina, Punjab.

Yasumaru CC, Xavier JG, Strefezzi RDF, Salles-Gomes COM. (2021). Intratumoral T-Lymphocyte Subsets in Canine Oral Melanoma and Their Association With Clinical and Histopathological Parameters. Veterinary Pathology.58(3):491-502.

11

Cryopreservation Methods of Primary Cultures and Cell Lines

Introduction

The protocol below describes the use of cell freezing methods involving an electric -80°C freezer for the cryopreservation of cell lines. ECACC routinely use a programmable rate-controlled freezer. This is the most reliable and reproducible way to freeze cells but as the cost of such equipment is beyond the majority of research laboratories the methods below are described in detail. If large numbers of cell cultures are regularly being frozen then a programmable rate-controlled freezer is recommended.

Materials

- Freeze medium (commonly 90% FBS, 10% DMSO or glycerol
- 70% (v/v) alcohol in sterile water
- PBS without Ca2+/Mg2+
- 0.05% trypsin/EDTA in HBSS, without Ca2+/Mg2+
- DMSO

Equipment

- Personal protective equipment (sterile gloves, laboratory coat)
- Full-face protective mask/visor
- Water bath set to 37°C

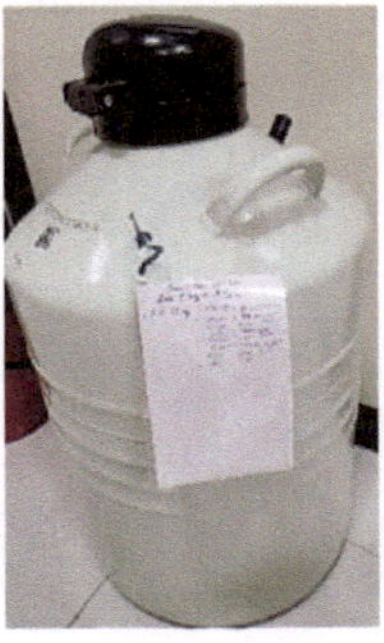

- Microbiological safety cabinet at appropriate containment level
- Centrifuge
- Hemocytometer (Bright-Line™ hemocytometer)
- Pre-labeled ampules/cryotubes
- Cell Freezing Device
- Liquid nitrogen (LN2) container filled with LN2

Procedure of freezing cells

1. For adherent cell lines harvest cells as close to 80 - 90% confluency as possible.
2. Remove spent media from culture vessel by aspiration (picture shown below)

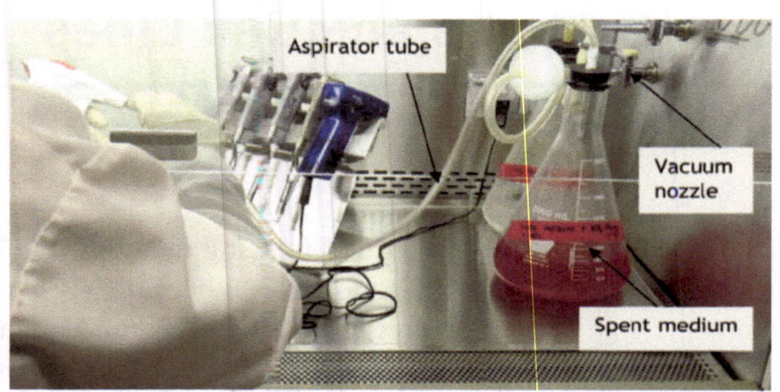

3. Wash the monolayer (2x with sterile PBS or salt solution free from Ca+2 and Mg+2 ions.
4. Bring adherent and semi adherent cells into suspension using trypsin/ EDTA as described previously. Add just sufficient volume of trypsin to cove cell monolayer. Incubate at 37 °C incubator for ~2 minutes.
5. Neutralize the trypsin by adding growth media.
6. Remove a small aliquot of cells (100-200µl) and perform a cell count. Ideally, the cell viability should be in excess of 90% to achieve a good recovery after freezing.
7. Centrifuge the remaining culture at 150 x g for 5 minutes.
8. Re-suspend cells at a concentration of 2-4x106 cells per ml in freeze medium.
9. Dispense 1 mL of cell suspension into labeled cryovials. Include the cell line name, passage number, lot number, cell concentration, and date on the labels.
10. Place ampoules inside a passive freezer e.g. Nalgene Mr. Frosty Freezing Container. Fill freezer with isopropyl alcohol and place at -80°C overnight.
11. Frozen ampoules should be transferred to the vapor phase of a liquid nitrogen and the locations recorded.

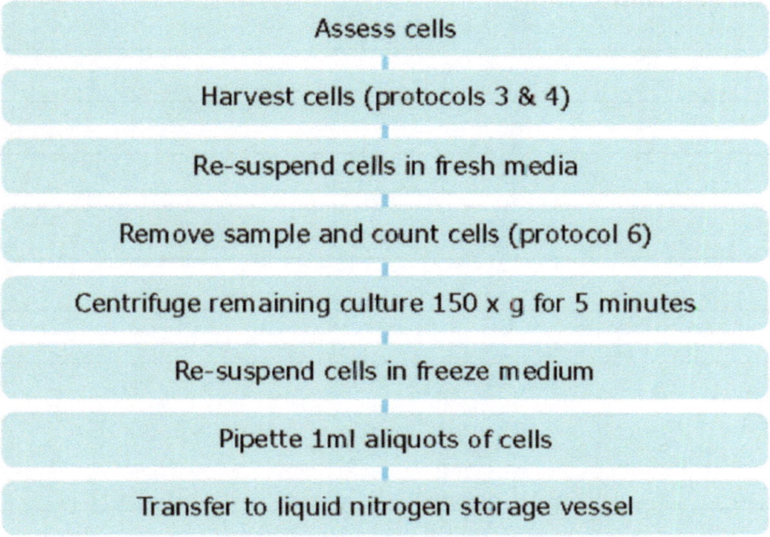

Assess cells

Harvest cells (protocols 3 & 4)

Re-suspend cells in fresh media

Remove sample and count cells (protocol 6)

Centrifuge remaining culture 150 x g for 5 minutes

Re-suspend cells in freeze medium

Pipette 1ml aliquots of cells

Transfer to liquid nitrogen storage vessel

Procedure of thawing of cells

1. Thaw one vial of frozen cells obtained from liquid nitrogen tank rapidly (< 1 minute) in a 37°C water bath.
2. Mix thawed cells slowly in pre-warmed growth medium and incubate.
3. Always use proper aseptic technique and work in biosafety cabinet.

Questions

Q1. Write down the applications of culturing PBMC.

Ans. ....

....

Q2. Why mitogens are added into the PBMC culture?

Ans. ....

....

Q3. Draw morphology of various types of PBMC.

Ans. ....

....

Further reading

Cell Culture Fundamentals: Cryopreservation and Storage of Cell Lines. ECACC Laboratory Handbook 4th Edition

Observation and Notes

12

Evaluating Cytopathic Effects of Viruses on Mammalian Cells

Introduction

Structural changes in host cells by the viral infection is referred as cytopathic effect (CPE) and the responsible virus is called cytopathogenic virus. As a virus hijacks a cell to make copies of itself, the cell goes through distinct changes in its chemistry and structure. These effects are easiest to see in lab-grown cell cultures. The severity of the damage caused by the virus depends on the specific virus itself, the type of cell it infects, how many viruses infect each cell (MOI), and other factors.

The early stages of this damage, called cytopathic effects (CPE), can be spotted directly under a microscope without any dyes or special treatments. While various types of CPE exist in living cells, some signs of viral infection, like inclusion bodies, require staining the cells to be visible. We can classify viruses based on how quickly they cause visible damage (cytopathic effect) in cell cultures. Viruses are typically considered slow if it takes 4-5 days to see this effect in cultures with a low starting dose of virus (MOI). On the other hand, viruses are considered fast if the damage shows up in just 1-2 days under the same conditions. It is important to note that at a high MOI all CPE can occur rapidly. So all decisions about rate of CPE appearance should be based on the lowest MOI that produces CPE.

Listed below are several general types of CPE

1. Total destruction
2. Subtotal destruction
3. Focal degeneration
4. Swelling and clumping
5. Foamy degeneration (vacuolization)
6. Cell fusion (syncytium)
7. Inclusion bodies

Protocols

General procedure for cultured animal cell infection

1. Carefully remove growth medium from cells.
2. Add 1 ml of maintenance medium (fresh medium with 2% serum) to each well.
3. Make 10-fold dilutions of your viruses in dilution tubes, using 1.8 ml of cold medium plus 0.2 ml of virus stock as diluent. Usually you will want to prepare a wide variety of dilutions of your virus such as 10-1 through 10-7.
4. Inoculate 0.1 ml of each dilution into one well of cells, leaving one uninoculated control well for each set of virus dilutions.
5. Incubate the cluster dish in a CO_2 incubator. Observe the dish daily in an inverted microscope to check for development of CPE.
6. Go to virology lab of College of Animal Biotechnology and follow ongoing experiments to observe CPE.
7. Draw the picture of your observation and label the types of CPE you observed in the lab.

Non-infected

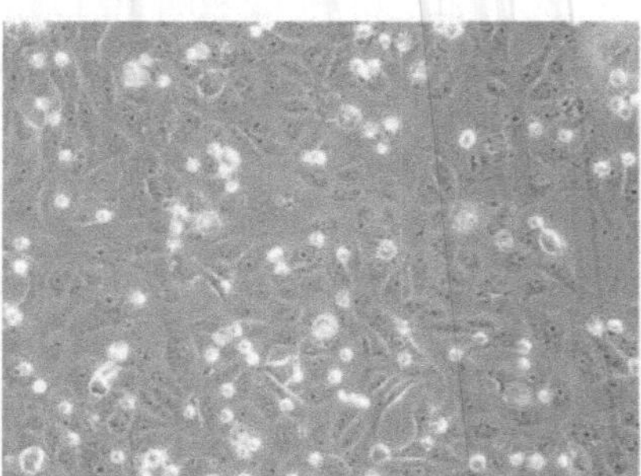

Virus infected cells showing CPE

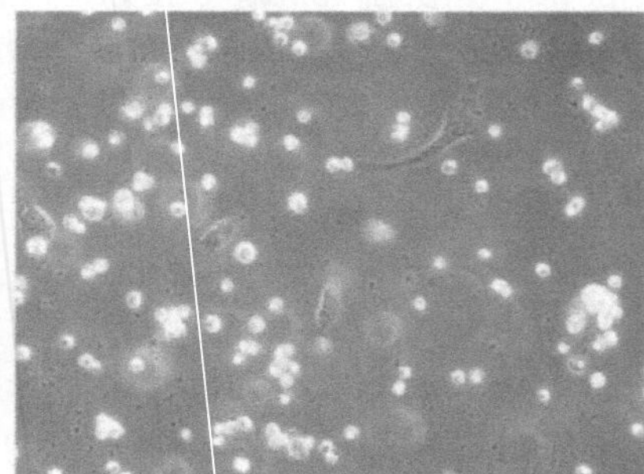

Question

Q1. How CPE is helpful in diagnosing the disease?

Ans. ....

....

Further reading

Knipes, D. M., C. E. Samuel, and P. Palese. 2001. Virus-host cell interactions, p.134–136. In D. M. Knipes andP. M. Howley (ed.),Fields virology, 4th ed. Lippincott Williams and Wilkins, Philadelphia, PA.

https://www.asmscience.org/content/education/protocol/protocol.2875 (Accessed on June 28 2021).

13

Virus Isolation in Cell Culture System

Principle

Isolating the virus in a laboratory is the gold standard for diagnosing many viral infections. Virus isolation is a highly specialised technique used to diagnose viral infections and produce a stock of viruses for research purposes. The isolation of the virus relies on the suitability of the sample, the ability of the cells to allow entry, the successful replication of the virus, and the release of viral particles from the cells. The choice of cell lines is dependent upon the suspected viral disease and the availability of such cell lines in the laboratory.

Materials required

Equipment

a) Biosafety cabinet level 2 (B2)
b) Carbon Dioxide (CO_2) Incubator (set to 37°C with a 5% CO_2 in air atmosphere)
c) Refrigerated centrifuge
d) Refrigerator
e) Inverted Microscope
f) Water bath
g) -86°C/-80°C deep freezer
h) Autoclave
i) Hot air oven
j) Filtration Assembly

Media and Materials

a) Cell culture flask 25 cc or 75 cc
b) Foetal bovine serum (FBS)
c) Growth media as per the type of cells (DMEM or EMEM with 10% FBS)
d) Maintenance media as per the type of cells (DMEM or EMEM with 2% FBS)

e) Glass Beaker (500 ml/1000 ml)
f) Sterile double distilled water
g) Plastic Serological Pipettes (2 ml, 5 ml, 10 ml and 25 ml).
h) 5 ml, 15 ml and 50 ml centrifuge tubes
i) Sodium Hypochlorite solution (10000 ppm)
j) Eppendorf tubes (1.5 ml and 2 ml)
k) Sterile micro pipette tips (1-10 µl, 10-200 µl, 100-1000 µl)
l) Micro pipettes (1-10 µl, 10-200 µl, 100-1000 µl)
m) 15 ml and 50 ml tube racks
n) 1.5 ml and 2 ml tube racks
o) Trypsin-Versene solution
p) Syringe Filter (0.45 µm pore size and 0.22 µm pore size)

Procedure

I. Seeding of cells

a) The cell culture procedures must be conducted within a Biosafety Level 2 (BSL 2) cabinet.

b) Sub-culture the cells in late log phase (2-3 days old monolayer) and scale up the cells in tissue culture flasks.

c) After sub-culturing, add pre-warmed (incubate at 37°C incubator for an hour) growth medium containing 10% foetal bovine serum (FBS). Determine the number of cells using a haemocytometer.

d) Seed the cell suspension to appropriate culture vessels to get a cell density of 5 x 10^5 cells / mL and add 20 ml of such stock in each 75 cc flask (1 x 10^7 cells / 75 cc flask).

e) Incubate cells at 37°C for 24-48 hours in a CO_2 Incubator with 5 % CO_2, when the cells attain 80-85% confluency they are ready for virus infection.

II. Preparation of inoculum

Clinical samples/tissues should be collected aseptically and rapidly transported to the laboratory in viral transport media (VTM). The sample for viral isolation should be immediately frozen at -80°C. Samples for viral isolation should be kept frozen continuously, avoiding freeze-thaw cycles that inactivate virus.

a) Take 10 g virus material (pieces of spleen or lymph nodes or tonsils or any other organ/tissue of infected/suspected animals)

b) Cut the tissue material into small pieces and grind it with sterile sand using a mortar and pestle inside a Biosafety cabinet II. The sand must be devoid of dust and sterile. Alternately, tissue homogenizer can also be used.

c) While grinding, add PBS with pH 7.4 so that the final volume becomes 50 ml, final concentration becomes 20%.

d) Distribute the suspension @ 2 ml in 5 ml tubes and freeze in -20°C/80°C deep freezer, overnight.

e) Take out one aliquot (2 ml) and thaw. Add 0.02 ml antibiotic solution (100 X solution,) and keep in ice for 30 min.

III. Infection of cells

a) Check the cells that have grown for 24-48 hours in a 75 cc flask or a 25 cc flask using an inverted compound microscope to assess their confluency. If the flasks contain a monolayer with a viability of 80-85%, it is appropriate for infection.

b) Discard the media from the flask and infect the cells with 200 µl of prepared tissue suspension in 25 cc flask or 600 µl in 75 cc flask and keep one uninfected flask as control.

c) Incubate at 37°C in a CO_2 incubator for one hour.

d) Take the flasks from the incubator, add the necessary growth media, and return them to the incubator.

e) Observe the cells for next 3-5 days for cytopathic effects (CPE). Example for CPE is described in Figure 1A and 1B.

f) Harvest the cell after 70% CPE or after 5 days

g) Freeze -thaw the harvested flask and infect the fresh cells in new flask with the cell culture lysate @ 200 µl of prepared tissue suspension in 25 cc flask or 600 µl in 75 cc flask.

h) Repeat the procedure for three passages.

i) After the completion of 3 blind passages, the cell lysate should be checked for the presence of suspected virus by using molecular technique such as PCR.

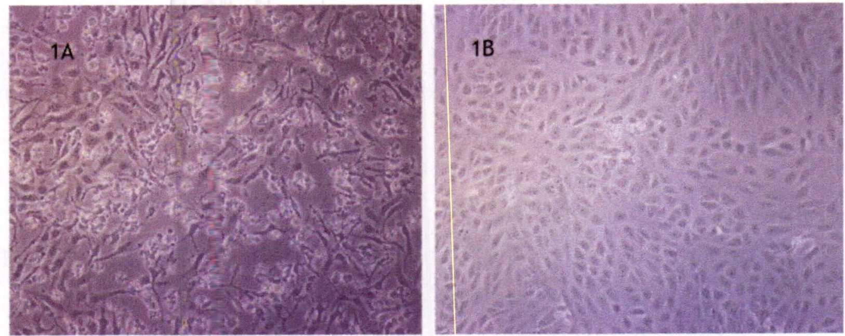

Figure 1A: Vero cells infected with Canine distemper virus 72 hours after infection on passage3, 1B: healthy Vero cells mock infection after 72 hours

Questions

Q1. How can primary cell cultures derived from different animal species be utilized to isolate and propagate specific viruses?

Ans. ....

....

....

....

Q2. Discuss the challenges associated with isolating and culturing fastidious animal viruses.

Ans. ....

....

....

....

Further reading

Fenner, F. J., Gibbs, E. P. J., Murphy, F. A., Rott, R., & Studdert, M. J. (Eds.). (2011). Veterinary Virology (4th ed.). Academic Press.

Mahy, B. W. J., & Van Regenmortel, M. H. V. (Eds.). (2009). Encyclopedia of Virology (3rd ed.). Academic Press.

Richman, D. D., Whitley, R. J., & Hayden, F. G. (Eds.). (2015). Clinical Virology (4th ed.). ASM Press.

14

One-Step Growth Curve of a Virus

Principle

The one-step growth experiment refers to a scientific methodology. A study aimed at observing the molecular events that take place during viral reproduction. This elucidates the fundamental features of the virus replication process and determines the optimal multiplicity of infection and timing for virus harvest. The initiation of modern bacteriophage research can be traced back to the development of the one-step growth curve by Max Delbrück and Emory Ellis in 1939, which utilised the Escherichia coli-T4 bacteriophage system. One step growth curve consists of three clearly defined phases 1. Latent period refers to the initial phase of a process where no significant changes or activities are observed. 2. Burst or Rise period is the subsequent phase characterised by a sudden increase virus particle. 3. Plateau period is the following phase where a stable and constant level virus replication.

Materials required

Equipment

a) Biosafety cabinet level 2 (B2)
b) Carbon Dioxide (CO_2) Incubator (set to 37°C with a 5% CO_2 in air atmosphere)
c) Refrigerated centrifuge
d) Refrigerator
e) Inverted Microscope
f) -86°C/-80°C deep freezer
g) Autoclave
h) Hot air oven
i) Filtration Assembly

Media and Materials

a) Cell culture flask 25 cc or 75 cc
b) Growth media as per the type of cells (DMEM or EMEM with 10% FBS)

c) Maintenance media as per the type of cells (DMEM or EMEM with 2% FBS)

d) Glass Beaker (500 ml/1000 ml)

e) Sterile double distilled water

f) Plastic Serological Pipettes (5 ml, 10 ml and 25 ml).

g) 15 ml and 50 ml centrifuge tubes

h) Sodium Hypochlorite solution (10000 ppm)

i) Eppendorf tubes (1.5 ml and 2 ml)

j) Sterile micro pipette tips(1-10 μl, 10-200 μl, 100-1000 μl)

k) Micro pipettes (1-10 μl, 10-200 μl, 100-1000 μl)

l) 15ml and 50ml tube racks

m) 1.5ml and 2ml tube racks

n) Trypsin-Versene solution

o) Syringe Filter (0.45 μm pore size and 0.22 μm pore size)

p) Phosphate buffered saline (PBS)

Procedure

a) Prepare 25cc flasks or 6-well plates (As per the requirement) that have cells with a confluency of over 85%.

b) Label the flasks or wells with the corresponding time intervals of 6, 12, 18, 24, 36, 48, 60, 72, 84, 96, 120, and 144 hours, indicating the time of harvest.

c) Prepare the virus inoculum for infection based on the desired multiplicity of infection (MOI), taking into consideration the specific requirements and characteristics of the virus to be used in the study.

d) Remove the media from the flasks/wells and rinse the cells with sterile PBS.

e) Inoculate the cells with the prepared virus inoculum, either in a flask or in wells.

f) Place the flask/plates in a CO_2 incubator (5% CO_2) at 37°C and allow it to incubate for 1 hour.

g) Remove the flasks/plates from the incubator and dispose of the inoculum. Wash the cells by rinsing them with sterile PBS.

h) Introduce 5ml of media into the flask or 1ml into the wells, then place the sample in a CO_2 incubator and incubate at a temperature of 37°C.

i) Harvest the cell flask based on their designated harvest time and store them in a deep freezer at a temperature of -80 degrees.

j) Perform two freeze-thaw cycles on the flask after the experiment is finished.

k) Perform individual virus titrations on each flask/well to determine the virus titre.

l) Calculate the virus titre by using Spearman-Karber method/Reed and Muench method/ M A Ramakrishna method and the virus titres will be represented in Tissue Culture Infective Dose 50 ($TCID_{50}$)

m) Generate a graph displaying the virus titre as a function of time.

n) Determine the optimal time for harvesting from the graph (Figure 1 displays a typical one-step growth curve).

Note: An experiment can be conducted to determine the optimal MOI by utilising varying MOI levels, such as low, medium, and high. The optimal MOI can be determined through one step growth curve analysis.

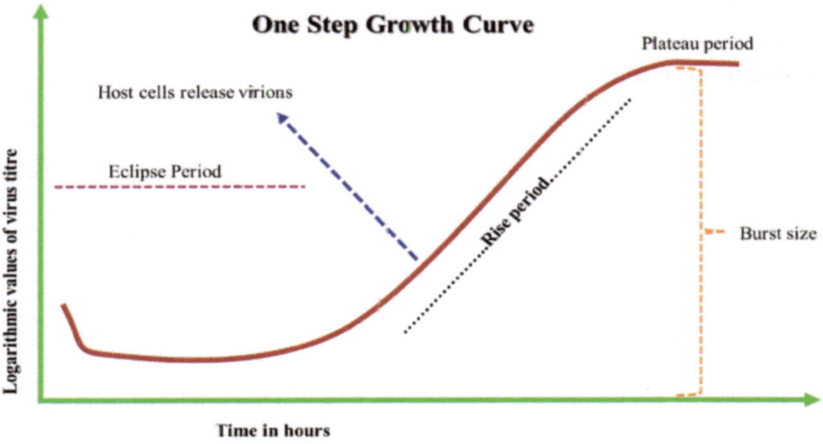

Figure 1: Displays a typical one-step growth curve

Questions

Q1. What are the reasons for representing viral growth with a one-step growth curve?

Ans. ....
....
....
....

Q2. Discuss the purpose of the one step growth curve?

Ans. ....
....
....
....

Q3. What is meant by "multiplication of infection"? and What is your understanding of MCI 10?

Ans. ....
....
....
....

Further reading

Cann, A. (Ed.). (1999). Virus culture: a practical approach (Vol. 208). Oxford University Press.

Kropinski, A. M. (2018). Practical advice on the one-step growth curve. Bacteriophages: Methods and Protocols, Volume 3, 41-47.

Richman, D. D., Whitley, R. J., & Hayden, F. G. (Eds.). (2015). Clinical Virology (4th ed.). ASM Press.

Observation and Notes

____

____

____

____

____

____

____

____

____

____

____

____

____

____

15

Production of Virus in Cell Culture

Principle

Unlike bacteria, which can be cultivated on an artificial nutrient medium, viruses rely on a living host cell for replication. Eukaryotic or prokaryotic host cells that have been infected with viruses can be used to produce viruses. Subsequently, the medium and cells can be collected and used as a virus source. The production of viruses usually consists of a two-step procedure: first, the growth of host cells, and then the production phase. Understanding the complex relationship between the host cell and virus is essential. Various crucial factors have been recognised, such as the concentration of host cells during infection and the multiplicity of infection (MOI).

Materials required

Equipment

a) Biosafety cabinet level 2 (B2)

b) Carbon Dioxide (CO_2) Incubator (set to 37°C with a 5% CO_2 in air atmosphere)

c) Refrigerated centrifuge

d) Refrigerator

e) Inverted Microscope

f) Micro pipettes(1-10 µl, 10-200 µl, 100-1000 µl)

g) -86°C/-80°C deep freezer

h) Autoclave

i) Hot air oven

j) Filtration Assembly

Media and Materials

a) Cell culture flask 25 cc or 75 cc

b) Growth media as per the type of cells (DMEM or EMEM with 10% FBS)

c) Maintenance media as per the type of cells (DMEM or EMEM with 2% FBS)
d) Glass Beaker (500 ml/1000 ml)
e) Sterile double distilled water
f) Plastic Serological Pipettes (5 ml, 10 ml and 25 ml).
g) 15 ml and 50 ml centrifuge tubes
h) Sodium Hypochlorite solution (10000 ppm)
i) Eppendorf tubes (1.5 ml and 2 ml)
j) Micro pipette tips(1-10 μl, 10-200 μl, 100-1000 μl)
k) 2 ml, 15 ml and 50 ml tube racks
l) 1.5 ml and 2 ml tube racks
m) Trypsin-Versene solution
n) Syringe Filter (0.45 μm pore size and 0.22 μm pore size)
o) Phosphate buffered saline (PBS)

Procedure

I. Adsorption method

a) Cultivate Vero/any other suitable cells in the late exponential phase (2–3 days old monolayer) and scale up the cell population in tissue culture flasks as needed.

b) Following the process of subculturing, add growth medium that has been pre-warmed and contains 10% foetal bovine serum (FBS).

c) Transfer the cell suspension into suitable culture flask in order to achieve a cell concentration of 5 x 10^5 cells per mL. Dispense 20 ml of this cell stock into each 75 cc flask, yielding an approximate cell count of 1 x 10^7 cells per flask. 5 millilitres should be added to each flask with a capacity of 25 cubic centimetres.

d) Incubate the cells at a temperature of 37°C for a duration of 24-48 hours until they reach 80–85% confluency, at which point they will be ready for virus infection.

e) Prepare the virus by adjusting the viral multiplicity of infection (MOI) according to the number of cells and viral titre. Store the virus on ice.

f) Take out the flasks from the incubator and dispose of the culture medium in a biosafety cabinet II.

g) Rinse the cells using sterile phosphate-buffered saline (PBS).

h) Inoculate the cell with the prepared virus and place it in an incubator at a temperature of 37 °C for a duration of one hour.

i) Remove the contents from the flask after one hour. Add the required media to the flask and incubate it at a temperature of 37°C for a period of five days or until cells show more than 80–85% cytopathic effects (CPE).

j) Harvest on the fifth day or when cells exhibit more than 80-85% cytopathic effects (CPE).

k) Place the flasks in a freezer set to a temperature of -80°C.

l) Take out the flasks from deep freezer and allow the flasks to thaw and then place them back in the freezer at a temperature of -80°C.

m) Perform the freeze-thaw process once again.

n) Perform a titration of the virus and then proceed to store it at -80°C.

II. Co culture method

a) Cultivate Vero cells until they reach the exponential phase, which typically occurs when the cells are 2-3 days old and have formed a monolayer.

b) After subculturing, suspend the cells in pre-warmed growth medium that is supplemented with 10% foetal bovine serum (FBS).

c) Place the cell suspension into appropriate culture vessels to attain a cell concentration of 5 x 10^5 cells per mL.

d) Transfer 20 ml of this cell stock into each 75 cc flask, resulting in an estimated cell count of 1 x 10^7 cells per flask. Each flask with a volume of 25 cc should receive an additional 5 millilitres. Adjust the viral multiplicity of infection (MOI) to match the number of cells and viral titre in order to prepare the virus. Store the virus on ice.

e) Inoculate the prepared virus into the cell and transfer the cell to an incubator set at a temperature of 37°C for a duration of five days or until the cells exhibit more than 80–85% cytopathic effects (CPE).

f) Harvest the cells on the fifth day or when they show more than 80–85% cytopathic effects (CPE).

g) Put the flasks inside a freezer adjusted to a temperature of -80°C.

h) Remove the flasks from the deep freezer and let them thaw at room temperature before returning them to the freezer at a temperature of -80°C.

i) Repeat the freeze-thaw process.

j) Conduct a virus titration and subsequently proceed with its storage.

Questions

Q1. What are the various techniques for virus production?

Ans. ....

....

....

....

Q2. How will you detect the growth of the virus in cell culture?

Ans. ....

....

....

....

Q3. Explain the reasons for cultivating viruses in cell culture.

Ans. ....

....

....

....

Further reading

Freshney, R. I. (2015). Culture of animal cells: a manual of basic technique and specialized applications. John Wiley & Sons.

Richman, D. D., Whitley, R. J., & Hayden, F. G. (Eds.). (2015). Clinical Virology (4th ed.). ASM Press.

Observation and Notes

____

____

____

____

____

____

____

____

____

____

16

Quantitative Analysis of Virus Titers through *in vitro* Assays

Principle

The quantification of infectious virus particles is commonly achieved through the utilisation of the Median Tissue Culture Infectious Dose (TCID50) assay. The assay functions through the introduction of a serial dilution of the virus sample into cells in a 96 well plate format. The chosen cell type is specifically intended to exhibit a cytopathic effect (CPE), which refers to observable changes in morphology or cell death resulting from viral infection. Following a period of incubation, the cells are examined for cytopathic effects or cell death, and each well is categorised as either infected or uninfected. The concentration at which 50% of the wells exhibit a cytopathic effect (CPE) is employed to determine the 50% tissue culture infectious dose ($TCID_{50}$) of the viral sample. The end point is the dose or dilution that infects or kills or shows CPE of the cells. The calculation can be performed using various mathematical techniques, such as the Spearman-Karber method or the Reed-Muench method. The virus titer is quantified as the number of tissue culture infectious dose 50 ($TCID_{50}$) per millilitre (ml). The non-cytopathic virus can be quantified by employing the fluorescent antibody test, immune-peroxidase assay, or other appropriate methodologies.

Procedure

Equipment

a) Biosafety cabinet level 2 (B2)
b) Carbon Dioxide (CO_2) Incubator (set to 37°C with a 5% CO_2 in air atmosphere)
c) Refrigerated centrifuge
d) Refrigerator
e) Inverted Microscope
f) Micro pipettes (1-10 µl, 10-200 µl, 100-1000 µl)

g) -86°C/-80°C deep freezer
h) Autoclave
i) Hot air oven
j) Filtration Assembly

Media and Materials

a) Cell culture flask 25 cc or 75 cc
b) 96 well cell culture plates
c) Growth media as per the type of cells (DMEM or EMEM with 10% FBS)
d) Maintenance media as per the type of cells (DMEM or EMEM with 2% FBS)
e) Glass Beaker (500 ml/1000 ml)
f) Sterile double distilled water
g) Plastic Serological Pipettes (5 ml, 10 ml and 25 ml).
h) 15ml and 50 ml centrifuge tubes
i) Sodium Hypochlorite solution (10000 ppm)
j) Eppendorf tubes (1.5 ml and 2 ml)
k) Micro pipette tips (1-10 µl, 10-200 µl, 100-1000 µl)
l) 2 ml, 15 ml and 50 ml tube racks
m) 1.5 ml and 2 ml tube racks
n) Trypsin-Versene solution
o) Syringe Filter (0.45 µm pore size and 0.22 µm pore size)
p) Phosphate buffered saline (PBS)
q) Positive virus control with known titre

Procedure

a) Perform virus titration in cell cultures in 96 well cell culture plate (Mark the plate as per the number of virus samples to be titrated, as shown in Figure 1).
b) Subculture cells and seed to the plate at a seeding density of $5x10^4$ cells per well (100^{-1} of cell stock of 5×10^5/ml).
c) Incubate the plate at 37°C with 5% CO_2 until the virus dilutions are prepared.
d) Prepare tenfold serial dilutions of virus (10^{-1} to 10^{-8}). Keep the virus dilutions on ice.

e) Transfer 100 µl of each virus dilution to the respective wells (3 wells for each dilution) in the plate keeping cell control.

f) For the control wells, add cell maintenance media and include a positive control virus for titration with a known titre to ensure accurate results.

g) Incubate the plate at 37°C and 5% CO_2 for 5 days

h) Remove the plate from the incubator and read the plate under an inverted microscope for CPE and record the results (An illustration of CPE-based virus titration at various dilutions is presented in Figure 2).

i) Calculate the virus titre by using Spearman-Karber method/Reed and Muench method/ and the virus titre will be represented in Tissue Culture Infective Dose 50 ($TCID_{50}$)

Note: For non-cytopathic viruses (e.g.: Rabies virus, Classical swine fever virus, Porcine circovirus, etc), the titration can be conducted using the aforementioned method, and the results can be obtained by performing a fluorescent antibody test (Figure 3A & 3B) or immune-peroxidase assay, or other suitable techniques.

	1	2	3	4	5	6	7	8	9	10	11	12
A	10^{-1}	10^{-2}	10^{-3}	10^{-4}	10^{-5}	10^{-6}	10^{-7}	10^{-8}			CC	CC
B	10^{-1}	10^{-2}	10^{-3}	10^{-4}	10^{-5}	10^{-6}	10^{-7}	10^{-8}			CC	CC
C	10^{-1}	10^{-2}	10^{-3}	10^{-4}	10^{-5}	10^{-6}	10^{-7}	10^{-8}			CC	CC
D	10^{-1}	10^{-2}	10^{-3}	10^{-4}	10^{-5}	10^{-6}	10^{-7}	10^{-8}				
E	10^{-1}	10^{-2}	10^{-3}	10^{-4}	10^{-5}	10^{-6}	10^{-7}	10^{-8}				
F	10^{-1}	10^{-2}	10^{-3}	10^{-4}	10^{-5}	10^{-6}	10^{-7}	10^{-8}				
G	10^{-1}	10^{-2}	10^{-3}	10^{-4}	10^{-5}							
H	10^{-1}	10^{-2}	10^{-3}	10^{-4}	10^{-5}							

	Sample 1
	Sample 2
	Positive control
	Cell control CC

Figure 1: Format for virus titration in 96 well plate

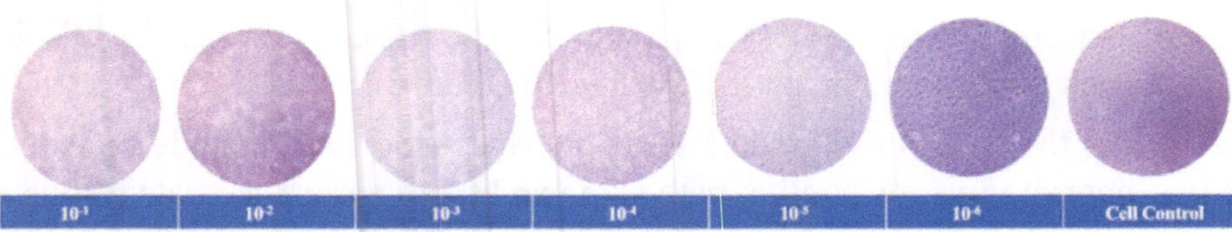

Figure 2: The virus sample was diluted from 10^{-1} to 10^{-6}, and cytopathic effects were observed from 10^{-1} to 10^{-5} on the 5th day post-infection. 10^{-6} and cell control are not displaying any cytopathic effects.

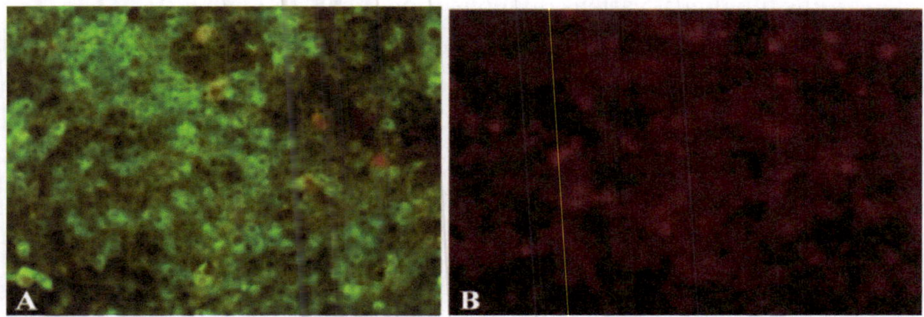

Figure 3A and 3B: PK15 cells infected with Classical swine fever virus 72 hours after infection detection of virus by fluorescent antibody test with Fluorescein isothiocyanate (FITC) dye and Evans Blue.

Formulas for the calculation of Virus titre

Any of these formulas or combinations can be used to calculate virus titre.

1. **Reed and Muench method:**

   $\log_{10}$ 50% end point dilution ($TCID_{50}$) = $\log_{10}$ of dilution showing a mortality next above 50% - (difference of logarithms × logarithm of dilution factor).

   Difference of logarithms = [(mortality at dilution next above 50%)-50%]/[(mortality next above 50%)-(mortality next below 50%)].

2. **Spearman-Karber method:**

   $\log_{10}$ 50% end point dilution = $-(x0 - d/2 + d\,\Sigma\, ri/ni)$

   x0 = $\log_{10}$ of the reciprocal of the highest dilution (lowest concentration) at which all wells are positive/ showing CPE

   d = $\log_{10}$ of the dilution factor

   ni = number of wells used in each individual dilution

   ri = number of wells shown CPE (out of ni).

   Summation is started at dilution x0.

3. **M A Ramakrishna Formula:**

   $\log_{10}$ 50% end point dilution = -[(total number of wells showing CPE/ number of wells inoculated per dilution) + 0.5] × log dilution factor.

Questions

Q1. What are the various techniques used for the determination of virus titre?

Ans. ....

Q2. How do you determine the titre of a virus?

Ans. ....

Q3. Discuss the differences between non-cytopathic and cytopathic virus titration.

Ans. ....

Further reading

Burleson, F. G., Chambers, T. M., & Wiedbrauk, D. L. (2014). Virology: a laboratory manual. Elsevier.

Cann, A. (Ed.). (1999). Virus culture: a practical approach (Vol. 208). Oxford University Press.

Freshney, R. I. (2015). Culture of animal cells: a manual of basic technique and specialized applications. John Wiley & Sons.

Richman, D. D., Whitley, R. J., & Hayden, F. G. (Eds.). (2015). Clinical Virology (4th ed.). ASM Press.

Observation and Notes

17

Hemadsorption (HA) Assay for Virus Titration

Principle

Red blood cell sticking (hemadsorption) can reveal the presence of certain viruses. This happens because these viruses, like influenza or measles, sprout proteins (hemagglutinin) on their infected host cells. These proteins act like Velcro, attracting red blood cells and causing them to cling to the surface. A hemadsorption assay uses this natural process to check for viruses. It's faster than waiting for the cells to show damage (cytopathic effect), allowing for earlier detection. Haemadsorbing viruses can be detected several days prior to the appearance of a cytopathic effect.

Equipment

a) Biosafety cabinet level 2 (B2)
b) Carbon Dioxide (CO_2) Incubator (set to 37°C with a 5% CO_2 in air atmosphere)
c) Refrigerated centrifuge
d) Refrigerator
e) Inverted Microscope
f) Micro pipettes (1-10 µl, 10-200 µl, 100-1000 µl)
g) -86°C/-80°C deep freezer
h) Autoclave
i) Hot air oven
j) Filtration Assembly

Media and Materials

a) Cell culture flask 25 cc or 75 cc
b) 24-well cell culture plate
c) Growth media as per the type of cells (DMEM or EMEM with 10% FBS)

d) Maintenance media as per the type of cells (DMEM or EMEM with 2% FBS)
e) Glass Beaker (500 ml/1000 ml)
f) Sterile double distilled water
g) Plastic Serological Pipettes (5 ml, 10 ml and 25 ml).
h) 15 ml and 50 ml centrifuge tubes
i) Sodium Hypochlorite solution (10000 ppm)
j) Eppendorf tubes (1.5 ml and 2 ml)
k) Sterile Micro pipette tips (1-10 µl, 10-200 µl, 100-1000 µl)
l) Micro pipettes (1-10 µl, 10-200 µl, 100-1000 µl)
m) 2 ml, 15 ml and 50 ml tube racks
n) 1.5 ml and 2 ml tube racks
o) Rypsin-Versene solution
p) Syringe Filter (0.45 µm pore size and 0.22 µm pore size)
q) Phosphate buffered saline(PBS)
r) Positive control
s) Blood for preparation of RBCs

Procedure

I. Preparation of RBCs

a) Collect blood from a guinea pig/chicken/human or another appropriate animal based on the specific virus being targeted for detection (For example, Canine parainfluenza virus: Guineapig RBCs, Influenza virus: Chicken RBCs).
b) Place red blood cells into a 15mL conical centrifuge tube.
c) Centrifuge at a speed of 500g for a duration of five minutes.
d) Remove the supernatant portion and the layer of white blood cells (Buffy coat).
e) Reconstitute red blood cells in chilled (2°C to 8°C) phosphate-buffered saline (PBS).
f) Wash RBCs with cold PBS until the liquid above the sediment is transparent, repeating this process two to three times.
g) Remove PBS by aspiration after the final wash.
h) Determine the remaining volume of cells by utilising the markings on the centrifuge tube. Include an adequate quantity of PBS to create a 10% concentration of cells(RBC stock solution).

i) Haemadsorption is performed by utilising a solution of red blood cells (RBCs) that has been prepared with a 0.4% concentration in a medium used for cell maintenance (RBC working solution).

II. Haemadsorption assay

a) Cultivate Vero or other appropriate cells in the late exponential phase (2–3 days old monolayer).

b) Perform subculturing, count the cells and then place them in 24-well plates.

c) Incubate the plates for a period of 24 to 48 hours.

d) Prepare virus dilutions and infect the 24-well plate containing cells with a confluency of over 85% (100 µl of virus) in duplicates, and add cell culture media to the control wells. Include a positive control virus for titration with a known titre to ensure accurate results.

e) The plate should be incubated in a CO_2 incubator set at a temperature of 37°C for a period of 3-5 days, depending on the specific virus being used.

f) Take out the flask after the completion of the incubation. Remove the maintenance medium from the infected wells and control cells in order to expose the cell monolayer.

g) Dispense 0.2 ml of the 0.4% red blood cell (RBC) suspension into each well.

h) Exercise caution to avoid cross-contamination of the wells.

i) Put the 24-well plate in a location with a temperature of 4°C for a duration of thirty minutes, and ensure that the erythrocyte suspension fully covers the monolayer of cells.

j) After incubation, immediately invert the plate to remove any red blood cells that are not adhering to the cell layer. Wash the plate with PBS gently.

k) Use an inverted light microscope with a 4× objective to inspect the cell culture tube for any red blood cells that are sticking to the monolayer. Immediately read all the wells after removing them from the refrigerator.

l) Negative: absence of hemadsorption of red blood cells Positive: The adherence of red blood cells (RBCs) to infected cells (Figure 1 and 2)

m) Calculate the virus titre by using Spearman-Karber method and Reed and Muench method

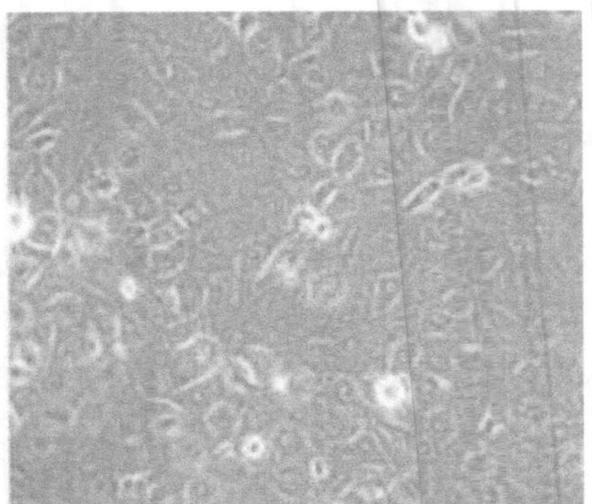

Figure 1: Healthy LLC-MMK2 cells (Picture Courtesy: Dr Mousumi Bora MVSc Research work Assam Agricultural University)

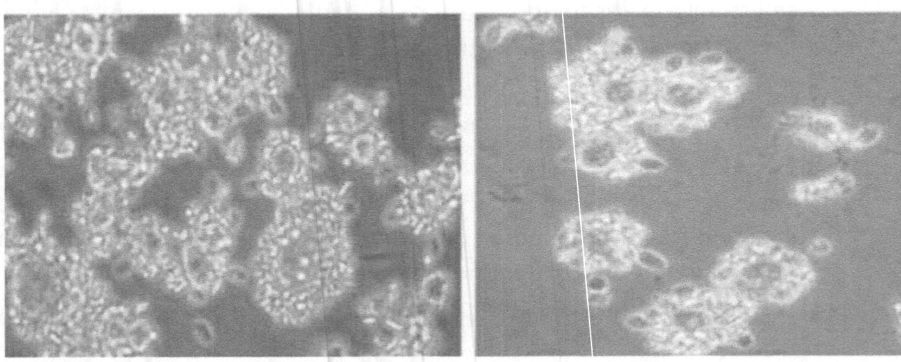

Figure 2: Infected LLC-MMK2 cells showing haemadsorption (Picture Courtesy: Dr Mousumi Bora, MVSc Research work, Assam Agricultural University)

Questions

Q1. What differentiates haemadsorption from haemagglutination?

Ans. ....

Q2. Explain the applications of haemadsorption.

Ans. ....

Q3. How does hemadsorption work?

Ans. ....

Further reading

Fenner, F. J., Gibbs, E. P. J., Murphy, F. A., Rott, R., & Studdert, M. J. (Eds.). (2011). Veterinary Virology (4th ed.). Academic Press.

Freshney, R. I. (2015). Culture of animal cells: a manual of basic technique and specialized applications. John Wiley & Sons.

Richman, D. D., Whitley, R. J., & Hayden, F. G. (Eds.). (2015). Clinical Virology (4th ed.). ASM Press.

18

Plaque Assay for Virus Detection and Quantification

Principle

The plaque assay was developed in order to enumerate and quantify the infectivity of bacteriophages. This technique has been subsequently adapted for application in animal virology and has proven to be a dependable method for quantifying the concentration of various viruses that have been adapted for growth in cell cultures. This assay is designed to quantify the capacity of a solitary infectious virus to create a visible area of cell death on a layer of cultured cells. A plaque forms during the viral infection cycle, in which the host cell dies after viral replication. Some viruses cause cell lysis during replication, while others cause minimal damage to the cells. The plaque assay is a technique used to measure the quantity of viruses that cause the destruction of their host cells, resulting in cytopathic effects (CPE). During a plaque assay, the host cells and virus are briefly co-cultured to enhance the virus's attachment and entry into the host cell. Subsequently, the concoction is introduced into a partially solidified agar medium. The thick agar is poured onto a layer of "bottom agar" which supplies ample nutrients for the host cell. After one round of virus replication, an infected cell will undergo lysis, leading to the release of multiple new viruses. Within the gelatinous substance, these newly released viruses can only infect neighbouring cells. After a subsequent round of replication, these neighbouring cells will undergo lysis. The cells that manage to avoid infection will continue to proliferate. After a period of 24 to 48 hours, plates that were not contaminated with a virus will display a consistent layer of cells. The plates infected with a multitude of viruses exhibit plaques. Quantifying viral titer is essential for assessing viral infectivity. The determination can be made either by conducting a plaque assay or by calculating the infectious dose.

Equipment

a) -86°C/-80°C deep freezer

b) Autoclave

c) Biosafety cabinet level 2 (B2)
d) Carbon Dioxide (CO_2) Incubator (set to 37°C with a 5% CO_2 in air atmosphere)
e) Filtration Assembly
f) Hot air oven
g) Inverted Microscope
h) Micro pipettes (1-10 µl, 10-200 µl, 100-1000 µl)
i) Refrigerated centrifuge
j) Refrigerator

Media and Materials

a) 1.5 ml and 2 ml tube racks
b) 15 ml and 50 ml centrifuge tubes
c) 2 ml, 15 ml and 50 ml tube racks
d) 96 well cell culture plates
e) Cell culture flask 25 cc or 75 cc
f) Eppendorf tubes (1.5 ml and 2 ml)
g) Glass Beaker (500 ml/1000 ml)
h) Growth media as per the type of cells (DMEM or EMEM with 10% FBS)
i) Maintenance media as per the type of cells (DMEM or EMEM with 2% FBS)
j) Micro pipettes (1-10 µl, 10-200 µl, 100-1000 µl)
k) Phosphate buffered saline (PBS)
l) Plastic Serological Pipettes (5 ml, 10 ml and 25 ml).
m) Positive virus control with known titre.
n) Sodium Hypochlorite solution (10000 ppm)
o) Sterile double distilled water
p) Sterile micro pipette tips (1-10 µl, 10-200 µl, 100-1000 µl)
q) Syringe Filter (0.45 µm pore size and 0.22 µm pore size)
r) Trypsin-Versene solution

Procedure

a) Grow the cell line in 96-well plates using the suitable growth medium. Ensure that the cells reach a state of confluency exceeding 90%.

b) Discard of the growth medium in the plates and wash with sterile Phosphate-Buffered Saline (PBS).

c) Prepare 10 serial dilutions of the virus sample, each one being ten times more diluted than the previous.

d) Administer the diluted virus into each well, utilising triplicate wells for each dilution.

e) Allow plates to incubate for 1-2 hours to enhance the uptake of viruses.

f) Remove the inoculum and wash using cell maintenance medium, if needed.

g) Add overlay medium containing low melting temperature agarose to the cells. Provide a sufficient period of time for the growth and development of an infection (5-7 days). Remove any overlay, if it exists.

h) Follow periodic checks of the cell monolayers to detect the presence of foci or plaques.

i) Calculate the number of plaques on plates that have 20 or more plaques. The multiplication of the mean plaque count per Petri dish by the virus dilution results in the quantity of plaque forming units (PFU) per unit volume of inoculum. The numerical quantity can be expressed as the viral titre.

Questions

Q1. Discuss the purpose of plaque assay.

Ans. ....

....

....

....

Q2. What is the function of agar in a plaque assay?

Ans. ....

....

....

....

Q3. How do you calculate plaque count?

Ans. ....

....

....

....

Further reading

Clokie, M. R., & Kropinski, A. (2009). Methods and protocols, volume 1: Isolation, characterization, and interactions. Methods in molecular biology". Humana press, 69-81.

Capes-Davis, A., & Freshney, R. I. (2021). Freshney's culture of animal cells: A manual of basic technique and specialized applications. John Wiley & Sons.

Cann, A. (Ed.). (1999). Virus culture: a practical approach (Vol. 208). Oxford University Press.

Fenner, F. J., Gibbs, E. P. J., Murphy, F. A., Rott, R., & Studdert, M. J. (Eds.). (2011). Veterinary Virology (4th ed.). Academic Press.

Ryu, W. S. (2016). Molecular virology of human pathogenic viruses. Academic Press.

Observation and Notes

19

Virus Neutralization Test

Principle

Neutralization refers to the process of reducing or nullifying the effect of a specific component. Neutralization is a process in which the effects of viral antigens or viruses reduce through Ag-Ab reactions. Neutralizing antibodies are a specific type of antibodies that play a role in neutralization. Not all antibodies generated within our body possess the capability to impede the effects of antigens. Only antibodies with the ability to neutralize can prevent the uptake of viral particles by cells and counteract their effects. The virus neutralization assay is a serological test utilized to identify the existence and extent of functional systemic antibodies that hinder the ability of a virus to cause infection.

Media and Materials

Equipment

a) Biosafety cabinet level 2 (B2)
b) Carbon Dioxide (CO_2) Incubator (set to 37°C with a 5% CO_2 in air atmosphere)
c) Refrigerated centrifuge
d) Refrigerator
e) Inverted Microscope
f) Micro pipettes (1-10 µl, 10-200 µl, 100-1000 µl)
g) -86°C/-80°C deep freezer
h) Autoclave
i) Hot air oven
j) Filtration Assembly

Media and Materials

a) Cell culture plate 96 well
b) Growth media as per the type of cells (DMEM or EMEM with 10% FBS)

c) Maintenance media as per the type of cells (DMEM or EMEM with 2% FBS)
d) Glass Beaker (500 ml/1000 ml)
e) Sterile double distilled water
f) Plastic Serological Pipettes (5 ml, 10 ml and 25 ml).
g) 15 ml and 50 ml centrifuge tubes
h) Sodium Hypochlorite solution (10000 ppm)
i) Eppendorf tubes (1.5 ml and 2 ml)
j) Sterile micro pipette tips (1-10 µl, 10-200 µl, 100-1000 µl)
k) Micro pipettes (1-10 µl, 10-200 µl, 100-1000 µl)
l) 2 ml, 15 ml and 50 ml tube racks
m) 1.5 ml and 2 ml tube racks
n) Trypsin - Versene solution
o) Syringe Filter (0.45 µm pore size and 0.22 µm pore size)
p) Phosphate buffered saline (PBS)
q) Positive and negative serum

Procedure

a) Grow the suitable cell line in 96-well plates using the appropriate growth medium. incubate the cells at 37°C in CO_2 incubator. Ensure that the cells reach a confluency level of over 90%.

b) Prepare serial 1:2 dilutions of the test serum and positive serum control in a microtiter plate or a similar format.

c) Dilute the virus to a concentration of 100 $TCID_{50}$ per 100 microliters.

d) Label a 96-well plate with duplicate dilutions for test serum, positive control, negative sera control, virus control, and cell control (Mark the plate as per the number of Serum samples to be titrated, as shown in Figure 1).

e) Add 100 microliters of test sera dilutions to the designated wells in a 96-well plate, and similarly add positive control serum and negative serum.

f) Dispense 100 µL of the virus dilution prepared in the previous step into the wells containing the serum. Thoroughly shake the plate to ensure proper mixing of the serum and virus. Place the plate in a CO_2 incubator at 37°C for 1 hour.

g) Take the plate containing cells, dispose of the media, and label the plate according to the serum-virus mix plate. Additionally, mark the plate for virus control and cell controls.

h) Take out the serum virus mix plate from the incubator once the incubation is finished, then add 100 µL of serum-virus mix to the wells containing cells as per the labelling.

i) Dispense 100 µL of virus dilution in duplicate into the virus control well and 100 microliters of maintenance media into the cell control wells.

j) Incubate the plate at 37°C for 4-5 days. Examine the plate under an inverted light microscope for cytopathic effects (CPE) after the incubation period.

k) When neutralising antibodies are found in the test sera, they attach to the virus, preventing infection of the cells. This results in no infection (No CPE) when the plates are examined using inverted microscopy. If antibodies are not present, cytopathic effects (CPE) will occur.

l) To calculate the virus titre, determine the highest dilution that does not result in cytopathic effects (no cell infection) and take the reciprocal of that value (An illustration of CPE-based virus titration at various dilutions is presented in Figure 2).

	1	2	3	4	5	6	7	8	9	10	11	12
A	1:2	1:4	1:8	1:16	1:32	1:64	1:128	1:256	1:512		CC	CC
B	1:2	1:4	1:8	1:16	1:32	1:64	1:128	1:256	1:512		CC	CC
C	1:2	1:4	1:8	1:16	1:32	1:64	1:128	1:256	1:512		CC	CC
D	1:2	1:4	1:8	1:16	1:32	1:64	1:128	1:256	1:512		VC	VC
E	1:2	1:4	1:8	1:16	1:32	1:64	1:128	1:256	1:512		VC	VC
F	1:2	1:4	1:8	1:16	1:32	1:64	1:128	1:256	1:512		VC	VC
G	1:2	1:4	1:8	1:16	1:32	1:64	1:128	1:256		1:2	1:4	1:8
H	1:2	1:4	1:8	1:16	1:32	1:64	1:128	1:256		1:2	1:4	1:8

Colour	Legend
	Sample 1
	Sample 2
	VC
	Cell control CC
	Positive control
	Negative control

Figure 1: Format for virus neutralization test in 96 well plate

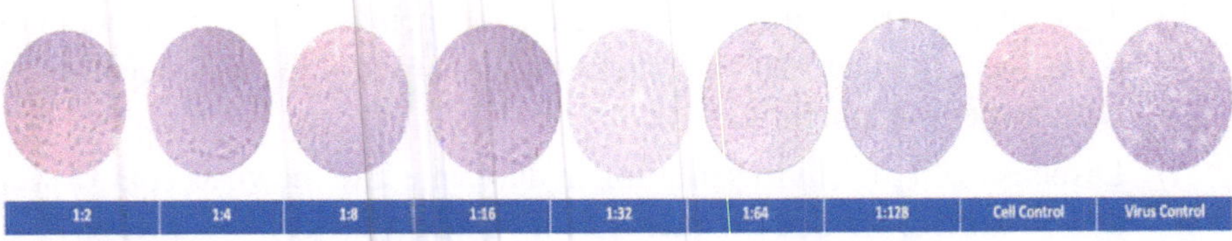

Figure 2: The serum sample was diluted from 1:2 to 1:128, and cytopathic effects were not observed up to 1:32 on the 5th day post-experiment. 1:64, 1:128 and virus control are displaying cytopathic effects

Questions

Q1. What are the mechanisms of virus neutralization?

Ans. ....
....
....
....

Q2. Explain the applications of virus neutralization test.

Ans. ....
....
....
....

Q3. Discuss the difference between neutralizing and non-neutralizing antibodies.

Ans. ....
....
....
....

Further reading

Cann, A. (Ed.). (1999). Virus culture: a practical approach (Vol. 208). Oxford University Press.

Capes-Davis, A., & Freshney, R. I. (2021). Freshney's culture of animal cells: A manual of basic technique and specialized applications. John Wiley & Sons.

Fenner, F. J., Gibbs, E. P. J., Murphy, F. A., Rott, R., & Studdert, M. J. (Eds.). (2011). Veterinary Virology (4th ed.). Academic Press.

Khare, R. (2020). Guide to clinical and diagnostic virology. John Wiley & Sons.

Ryu, W. S. (2016). Molecular virology of human pathogenic viruses. Academic Press.

20

Techniques for Establishing and Maintaining Cell Suspension Cultures

Principle

The majority of cell lines are cultivated as monolayer adherent cells, meaning they only grow on the surfaces of culture vessels. However, there are particular cells that are non-adhesive and do not rely on external support for their growth, referred to as suspension cells. A suspension culture, is a cell culture where individual cells or small cell clusters grow and multiply in a stirred growth medium, forming a suspension. Subculturing, also known as passaging cells, involves transferring cells from a previous culture into fresh growth medium to allow the cell line to continue growing. Suspension cultures involve the suspension of cells in a liquid medium and agitation at a specific speed to ensure that the cells are uniformly exposed to nutrient media from every angle. This method prevents cell aggregation and enhances nutrient absorption, thereby promoting cell growth. The maintenance of these cultures is achieved by transferring the cells from the early stationary phase to a fresh medium through subculturing. During the incubation period, there is an increase in both cell division and cell enlargement, leading to rapid cellular growth. The viability of cells in the suspension decreases after the stationary phase due to the exhaustion of certain factors or the accumulation of toxic substances in the medium. This decrease in the viability of the cells further reduces the growth rate of the entire culture. Next, a portion of the cell suspension is added to the freshly prepared medium with the same composition.

Equipment

a) Biosafety cabinet level 2 (B2)
b) Carbon Dioxide (CO_2) Incubator (set to 37°C with a 5% CO_2 in air atmosphere)
c) Refrigerated centrifuge
d) Refrigerator
e) Inverted Microscope

f) Micro pipettes (1-10 µl, 10-200 µl, 100-1000 µl)
g) -86°C/-80°C deep freezer
h) Autoclave
i) Hot air oven
j) Filtration Assembly

Media and Materials

a) Cell culture Flasks 25 cc and 75 cc
b) Media as per the type of cells (example SFM 900)
c) Glass Beaker (500 ml/1000 ml)
d) Sterile double distilled water
e) Plastic Serological Pipettes (5 ml, 10 ml and 25 ml).
f) 15 ml and 50 ml centrifuge tubes
g) Sodium Hypochlorite solution (10000 ppm)
h) Eppendorf tubes (1.5 ml and 2 ml)
i) Sterile micro pipette tips (1-10 µl, 10-200 µl, 100-1000 µl)
j) Micro pipettes (1-10 µl, 10-200 µl, 100-1000 µl)
k) 2 ml, 15 ml and 50 ml tube racks
l) 1.5 ml and 2 ml tube racks
m) Syringe Filter (0.45 µm pore size and 0.22 µm pore size)
n) Phosphate buffered saline (PBS)

Procedure

a) Culture the selected cells in an appropriate medium under optimal environmental conditions.
b) Once the cells are in the log-phase of growth but before reaching confluency, retrieve the flask from the incubator and extract a small sample from the culture flask using a sterile pipette in a Biosafety level II cabinet. If cells have settled, swirl the flask to distribute them evenly in the medium before sampling.
c) Determine the total cell count and viability (by Trypan Blue Dye Exclusion Cell Viability Assay) percentage using a haemocytometer from the sample.
d) Centrifuge the cell suspension in a sterile centrifuge tube to separate the cells from the media at 5000 rpm for 5 minutes at 4°C.
e) Discard the media and add the appropriate amount of fresh media (depending on the cell count) for splitting the cells into different flasks according to the cell count.

e) Transfer the cells to the new flasks and place them in an incubator at 37°C for 2-3 days.

f) Once the cells have been ready for subculture, the same procedure can be repeated.

g) The cells can be utilised for various purposes such as virus production, protein expression, and molecular pathogenesis studies.

h) The cells can be preserved for an extended period by cryopreserving them in liquid nitrogen (LN_2).

Note:

a) Large-scale suspension culture cell cultivation is possible using spinner flasks.

b) When using spinner flasks, it is recommended to use 30 ml of media for a 100 ml flask, 80 ml for a 250 ml flask, and 200 ml for a 500 ml flask.

c) The ideal temperature for culturing mammalian cells is 37°C, while for insect cells it is 27°C.

d) The choice between serum-containing media and serum-free media depends on the nature of cells.

e) Typically, serum-free media is utilised in insect cell culture. It is not suggested to use CO_2 incubator for insect cell culture.

f) An incubator maintained at a temperature of 27°C is adequate for culturing insect cells.

g) It is necessary to perform daily microscopic examinations of cells to verify their health and conformity to expected growth patterns. Suspension cells should exhibit a spherical shape and possess a plump appearance, causing them to scatter light around their membrane (Figure 1). Certain suspension cells may exhibit a clumping morphology.

h) There is no need to change the medium when cultivating cells in suspension. For regular subculturing, it is necessary to remove the cell suspension and add enough medium to dilute the culture to the desired density.

i) Suspension cells exhibit a high sensitivity to physical shearing. It is important to exercise caution in order to prevent any harm to the cells.

j) Preserve the spent media for analysis if the cell line is a hybridoma or another cell line that produces a compound such as a recombinant protein or growth factor.

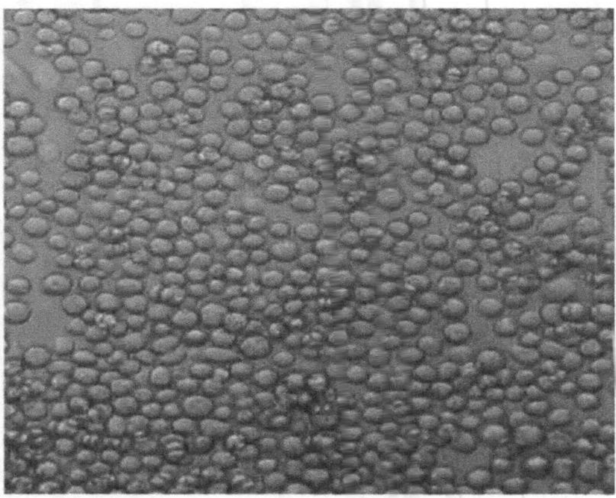

Figure 1. Displays suspension cells with more than 80% confluency

Questions

Q1. How do suspension culture and monolayer culture differ from one another?

Ans. ....
....
....
....

Q2. Discuss the uses of suspension culture in virology.

Ans. ....
....
....
....

Q3. What are some examples of suspension cell cultures?

Ans. ....
....
....
....

Further reading

Capes-Davis, A., & Freshney, R. I. (2021). Freshney's culture of animal cells: A manual of basic technique and specialized applications. John Wiley & Sons.

Maramorosch, K. (Ed.). (2012). Practical tissue culture applications. Elsevier.

Meyer, H. P., & Schmidhalter, D. (2014). Industrial scale suspension culture of living cells. John Wiley & Sons.

Paul Jr, F. (Ed.). (2012). Tissue culture: methods and applications. Elsevier.

21

Development of Cell Lines Through Virus-Mediated Techniques

Principle

Adenoviral, oncoretroviral, and lentiviral vectors are commonly utilised for gene delivery in mammalian cell culture and in vivo. Additional important examples of viral gene transfer involve baculovirus and vaccinia virus vectors. Stable cell lines produced with lentivirus technology provide consistent, long-term protein expression compared to the short-term expression achieved through transient transfection. This chapter focuses on cell transformation by lentiviruses.

Equipment

a) Biosafety cabinet level 2 (B2)
b) Carbon Dioxide (CO_2) Incubator (set to 37°C with a 5% CO_2 in air atmosphere)
c) Refrigerated centrifuge
d) Refrigerator
e) Inverted Microscope
f) Micro pipettes (1-10 μl, 10-200 μl, 100-1000 μl)
g) -86°C/-80°C deep freezer
h) Autoclave
i) Hot air oven
j) Filtration Assembly

Media and Materials

a) Cell culture flask 25 cc or 10 cm cell culture flask
b) 2 ml, 15 ml and 50 ml tube racks
c) Trypsin-Versene solution
d) Phosphate buffered saline (PBS)

e) 293T cells
f) Polybrene
g) Growth media as per the type of cells (DMEM or EMEM or suitable media with 10% FBS)
h) Maintenance media as per the type of cells (DMEM or EMEM or suitable media with 2% FBS)
i) Glass Beaker (500 ml/1000 ml)
j) Sterile double distilled water
k) Plastic Serological Pipettes (5 ml, 10 ml and 25 ml).
l) 15ml and 50 ml centrifuge tubes
m) Sodium Hypochlorite solution (10000 ppm)
n) Sterile micro pipette tips (1-10 µl, 10-200 µl, 100-1000 µl)
o) Micro pipettes (1-10 µl, 10-200 µl, 100-1000 µl)
p) Syringe Filter (0.45 µm pore size and 0.22 µm pore size)
q) 1.5 ml and 2 ml tube racks
r) Eppendorf tubes (1.5 ml and 2 ml)

Procedure

a) Seed approximately 2-3 x 10^5 293T cells in a 10 Cm tissue culture plate or 25 cubic centimetre flask and incubate at 37°C for 1-2 days
b) 293T cells should be around 70-80% confluent before the transfection.
c) Prepare the transfection mix in a sterile 1.5 ml Eppendorf tube according to the manufacturer's instructions provided in the kit containing 1.5µg of target plasmid.
d) Allow the mixture to incubate for approximately 20 minutes at room temperature (25°C).
e) Remove the medium from 293T cells plate/flask using disposable sterile pipette and replace it with approximately 9 ml of fresh Growth DMEM medium.
f) After incubating the transfection mix for 20 minutes, transfer the mix to the 293T cells using a 200µl micropipette.
g) Place the plate/Flask in a CO2 incubator at 37 °C for approximately 48 hours.
h) Transfer the supernatant/media from the 293T plate/flask into a sterile 15 ml tube using a 10 ml disposable sterile pipette.

i) Add an additional 10 ml of fresh DMEM media to the plate and return it to the incubator.

j) Transfer the gathered supernatant into a 10 ml syringe and filter it using a 0.45 µm PVDF filter within the Biosafety cabinet level II (A2/B2). This step will lead to the collection of filtered viral supernatant free of 293T cells.

k) Repeat steps g to j to obtain a larger quantity of virus. If the target cells are not prepared, store the lentivirus in liquid nitrogen storage facility or a -80 °C deep freezer.

l) Ensure thorough disinfection and proper disposal of the plate/flask utilised for virus production.

m) Seed approximately 1-2 x 10^5 target cells into two 10Cm plate/25cc tissue culture flask. One flask/plate is for virus infection and the other is for cell control.

n) The target cells should be approximately 40-50% confluent before transduction with the virus.

o) Once the cells are prepared for infection, make a polybrene medium mixture. The ultimate concentration of polybrene should be 8µg/ml. Prepare a 10x polybrene solution by combining 1 ml of medium with 10 µl of 8 mg/ml Polybrene.

p) Remove the medium from the target cells using a disposable sterile pipette. Add 1 ml of polybrene medium mix and allow it to incubate for approximately 1-2 minutes.

q) Add a medium without any additives in the cell control plate.

r) Place the plate in a 37°C cell culture incubator for approximately 48 hours.

s) Add the selection antibiotic to both flasks at the specified concentration.

t) Typically, fresh medium containing antibiotics is replenished every two days until all cells in the control flask have killed.

u) Prior to conducting additional experiments, ensure to cryopreserve approximately one-fourth of the cells in cryopreservation media containing 10% DMSO , then store them in liquid nitrogen.

v) Passage the remaining cells for RNA/protein isolation to confirm the consistent expression of the protein of interest.

Questions

Q1. What is the mechanism by which viruses induce cellular transformation?

Ans. ....

....

....

....

Q2. What is the virus mediated gene transfer method?

Ans. ....

....

....

....

Q3. Explain the various types of viruses that induce cell transformation.

Ans. ....

....

....

....

Further reading

Al-Rubeai, M. (Ed.). (2009). Cell Line Development (Vol. 6). Springer Science & Business Media.

Capes-Davis, A., & Freshney, R. I. (2021). Freshney's culture of animal cells: A manual of basic technique and specialized applications. John Wiley & Sons.

Niazi, S., & Lokesh, S. (2022). Biopharmaceutical Manufacturing, Volume 2: Unit processes. IOP Publishing.

Pontén, J., & Pontén, J. (1971). Spontaneous and virus induced trans-formation in cell culture (pp. 1-253). Springer Vienna.

Observation and Notes

____

____

____

____

____

____

____

____

22

Cell Viability and Cell Proliferation Assay

Principle

Cell viability refers to the quantity of healthy cells in a sample, while cell proliferation is a crucial indicator for comprehending the mechanisms and pathways involved in the survival of cells or death following exposure to toxic substances. The methods used to assess viability are typically the same as those used to detect cell proliferation. Cell cytotoxicity and proliferation assays are commonly employed in drug evaluations to determine if the test compounds impact cell proliferation or exhibit direct cytotoxic effects. Cell viability can be determined by calculating the ratio of total live cells to the total number of cells (live and dead). Staining also helps to observe the general cell structure. Cell proliferation assay by MTT. Metabolically active cells reduce the yellow tetrazolium MTT (3-(4, 5-dimethylthiazolyl-2)-2, 5-diphenyltetrazolium bromide) through dehydrogenase enzymes to produce NADH and NADPH. The purple formazan produced inside the cell can be dissolved and measured using spectrophotometry. The MTT Cell Proliferation Assay quantifies cell proliferation rate and detects decreases in cell viability caused by metabolic events leading to apoptosis or necrosis. The MTT Reagent produces minimal absorbance values when no cells are present.

Equipment

a) -86°C/-80°C deep freezer

b) Autoclave

c) Biosafety cabinet level 2 (B2)

d) Carbon Dioxide (CO2) Incubator (set to 37°C with a 5% CO2 in air atmosphere)

e). Filtration Assembly

f) Hot air ovenMedia and Materials

g) Inverted Microscope

h) Micro pipettes (1-10 µl, 10-200 µl, 100-1000 µl)

i) Refrigerated centrifuge

j) Refrigerator

Media and Materials

a) 1.5 ml and 2 ml tube racks
b) 15 ml and 50 ml centrifuge tubes
c) 2 ml, 15 ml and 50 ml tube racks
d) 6-well or 12-well culture tissue culture flask
e) Cell culture flask 25 CM^2 or 75 CM^2
f) Eppendorf tubes (1.5 ml and 2 ml)
g) Glass Beaker (500 ml/1000 ml)
h) Growth media as per the type of cells (DMEM or EMEM with 10% FBS)
i) Maintenance media as per the type of cells (DMEM or EMEM with 2% FBS)
j) Micro pipette tips (1-10 µl, 10-200 µl, 100-1000 µl)
k) MTT Reagent
l) Phosphate buffered saline (PBS)
m) Plastic Serological Pipettes (5 ml, 10 ml and 25 ml).
n) Sodium Hypochlorite solution (10000 ppm)
o) Sterile double distilled water
p) Syringe Filter (0.45 µm pore size and 0.22 µm pore size)
q) Trypsin-Versene solution

Procedure

a) Incubate cells for 48 hours at 37°C and 5% CO_2.

b) Inoculate in 100 µl of culture medium with different concentrations of drug to be tested and cell control wells. Add positive control to the positive control well.

c) Seed the suitable cells at a density of 4×10^5 cells per well of 6 well plates (two set of plates one for cell viability and other for cell proliferation assay).

d) Take out the plate and perform trypsinization on the cells, then take 100ul of the cell suspension

Prepare the cell dilution

e) Mix 10 µl of your cell suspension with 100 µl of Trypan blue solution in a small (1.5 ml) micro centrifuge tube.

f) Use a micropipette capable of handling 200 microliters to gently mix the solution several times.

Prime the micropipette

g) Discard the first few drops of the mixed cell suspension after drawing it into the pipette. This ensures you're not using any leftover solution from previous steps.

Fill the hemocytometer chamber

h) Carefully touch the tip of the pipette to the edge of the hemocytometer chamber.

i) Let the diluted cell suspension flow into the chamber by capillary action. The chamber should only be filled up to the level of the grooves, not overflowing.

Observe under the microscope

j) Focus on the grid lines of the hemocytometer.

k) Switch your microscope to the 10x objective lens.

Identify the counting area

l) Locate the specific set of 16 small squares in the corner of the hemocytometer, typically marked by a circle in diagrams (shown here). This is the area you'll use to count your cells.

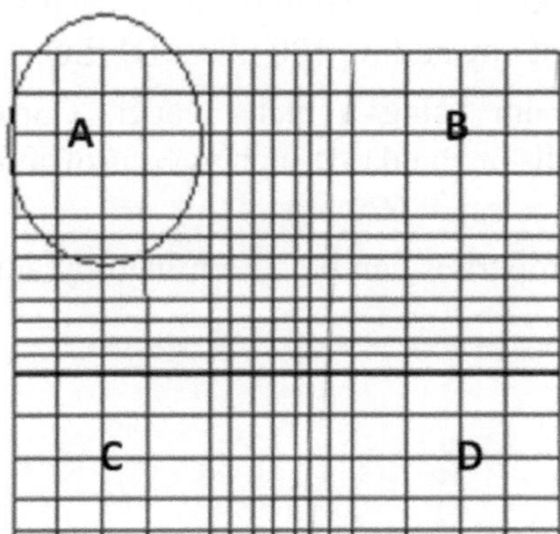

Count the cells in a defined area: Use a hand counter to tally the number of cells visible within a specific grid of 16 squares (area A) on the hemocytometer.

Repeat for all counting areas: Move the hemocytometer to view and count the cells in three additional sets of 16 squares (areas B, C, and D).

Factor in the hemocytometer design: The hemocytometer is designed such that the average number of cells counted in one set of 16 squares (A, B, C, or D) represents the actual number of cells per milliliter of cell suspension diluted by a factor of 10,000 (10^4).

Calculating the Total Cell Count

- To determine the final cell concentration per milliliter (ml), you'll need to:
- Add the cell counts from all four areas (A + B + C + D).
- Multiply this sum by the dilution factor used to prepare the cell suspension.
- Multiply the final result by 10,000 (10^4) to account for the hemocytometer's design.

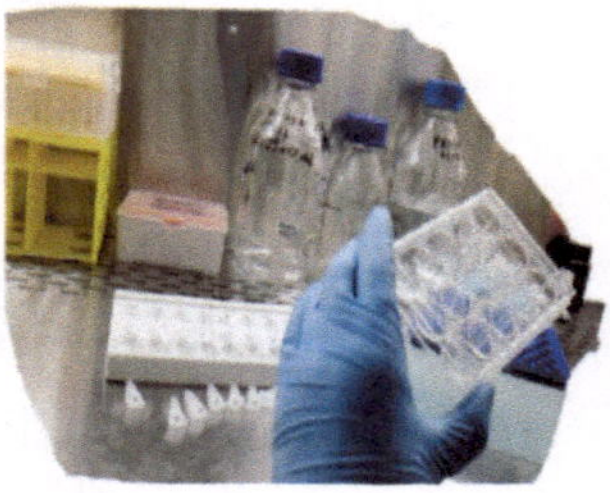

Cell proliferation assay by MTT

a) Prepare 5 mg/mL solution in PBS/ cell culture media. Mix by vortexing or sonication.

b) Sterilise the solution by filtering it. MTT solution can be stored at -20°C for a minimum of 6 months. Avoid storing at 4°C for an extended period

c) Remove the old media and add 100 microliters of MTT solution to each well. Start incubation for 4 hours.

d) After 4 hours, carefully remove the MTT solution from each well and dissolve the formazan precipitate by adding 100 uL of DMSO.

e) Measure absorbance at 570 nm using a plate reader. Correct for background by using blank wells or standards such as control and blank. The reference wavelength can be set at 620 nm.

f) The drug's cell proliferative properties can be determined by analysing the OD values.

Questions

Q1. What factors influence the results from the MTT assay?

Ans. ....

....

....

....

Q2. What methods are used to measure cytotoxicity?

Ans. ....

....

....

....

Q3. Explain various techniques for assessing cell viability.

Ans. ....

....

....

....

Further reading

Bhatt, T. K., & Nimesh, S. (Eds.). (2021). The design and development of novel drugs and vaccines: Principles and protocols. Academic Press.

Capes-Davis, A., & Freshney, R. I. (2021). Freshney's culture of animal cells: A manual of basic technique and specialized applications. John Wiley & Sons.

Castell, J. V., & Gmez-Lechn, M. J. (Eds.). (1996). In vitro methods in pharmaceutical research. Elsevier.

Moravec, R. A., ABS, A. L. N., Duellman, S., Benink, H. A., Worzella, T. J., & Minor, L. Cell Viability Assays.

Tekade, R. K. (Ed.). (2022). Pharmacokinetics and Toxicokinetic Considerations-Vol II. Academic Press.

Observation and Notes

Glossary

293T (HEK 293T) is a human cell line. It is frequently utilised in biological research to synthesise proteins and generate recombinant retroviruses.

Antigen recall assay: It is an antigen specific system assesses the activity of potential immune checkpoint inhibitors (ICI) candidates, in a relatively physiologically relevant manner.

Apoptosis: is the process of programmed cell death. It is opposite to the cell proliferation.

BSS: BSS, which stands for Balanced Salt Solution, plays a critical role in keeping cells happy and healthy in cell culture experiments. BSS is formulated to closely resemble the natural fluids surrounding cells within the body. This includes having the right balance of salts (like sodium, potassium, calcium, and magnesium) at a concentration similar to what cells are used to. Cell line is a specific group of cells that can be cultured for a long time while maintaining consistent characteristics and functions. Cell proliferation is the cellular process of growth and division resulting in the production of two daughter cells. Confluency is the percentage of the culture flask's or plate's surface area that is occupied by cells adhering to it.

Confluency: It is the percentage of the surface area of a culture dish that is covered by adherent cells.

Cryopreservation: is a process that preserves cells or tissues, or any other biological materials (like DNA, RNA, proteins) at very low temperatures (like liquid nitorgen has a temperature of -196°C.

Cytokines: Cytokines the the signalling molecules secreted by various immune cells.

Cytopathic effect (CPE): CPE refers to the structural changes in host cells caused by virions. Viruses hijack the host cell's machinery to replicate themselves. This process disrupts the normal functions of the cell and ultimately leads to its death. Some common examples of CPE include:

- Cell rounding: Healthy cells are typically flat and spread out. Virus infection can cause them to become round or balloon-like.

- Syncytia formation: Some viruses can cause adjacent infected cells to fuse together, forming giant multi-nucleated cells called syncytia.
- Inclusion bodies: These are abnormal structures that appear within the cell's cytoplasm or nucleus due to viral replication.
- Cell lysis: This is the bursting or breaking open of the cell, releasing the newly formed viruses into the environment to infect other cells.

Cytotoxicity: Refers to the degree of toxicity a substance has on cells. A cytotoxic compound can induce cell damage or death, either by necrosis or apoptosis.

EDTA: Ethylenediaminetetraacetic acid (EDTA) is an aminopolycarboxylic acid, is a chelating agent.

FBS: Fetal bovine serum is the liquid fraction of clotted blood from fetal calves, depleted of cells, fibrin and clotting factors, but containing a large number of nutritional and macromolecular factors essential for cell growth

Ficoll: Ficoll substance is a neutral, complex, large, water-loving sugar molecule which easily mixes in watery solutions. Haemadsorption: Red blood cells adhering to the surface of cells infected with specific viruses like herpes simplex, influenza virus, parainfluenza virus etc. Viral proteins are inserted to aid in the virus's release from the cell membrane, which in turn helps red blood cells bind to the infected cells. This method can utilise red blood cells from various species such as human O, chicken, or guinea pig. Hemocytometer, also referred to as a cell counting chamber, is a device utilised for manual cell counting.

Hemocytometer: It is thick glass slide with grid of perpendicular lines used to count cell numbers. The hemocytometer is divided into 9 major squares of 1 mm x 1 mm size. The four coner squares (red squares) are further subdivided into 4 x 4 grids. The height of the chamber formed with the cover glass is 0.1 mm, so a 1 mm x 1 mm x 0.1 mm chamber has a volume of 0.1 mm^3 or 10-4 ml.

Histopaque: Histopaque is a solution of polysucrose and sodium diatrizoate, used along with Ficoll for rapid recovery of viable lymphocytes and other mononuclear cells from whole blood Multiplicity of infection (MOI) indicates the quantity of virions introduced per cell during infection.

Multiplicity of infection (MOI): number of virions that are added per cell during infection. If one million virions are added to one million cells, the MOI is one.

MTT assay: It is a colorimetric assay for assessing cell metabolic activity. It is a widely used assay in cell biology research to measure cell viability, proliferation, and cytotoxicity

One-step growth curve experiment is a method used to observe molecular events that take place during the replication of a virus. It elucidates the fundamental essence of the virus replication process.

Plaque assay: An assay used for virus isolation, purification, and quantification of viral titers. Polybrene: Transfection Reagent provides effective gene delivery into mammalian cells by utilising retroviral vectors for infection.

PPE: PPE kit stands for Personal Protective Equipment kit. It's a set of gear worn to protect the wearer from contamination or injury. The specific items in a PPE kit can vary depending on the situation, but common components include:

- Masks or respirators: These cover the mouth and nose and filter out particles that could be harmful if inhaled.
- Gloves: Protect hands from contact with contaminants.
- Gowns: Disposable garments that cover the wearer's clothes and skin.
- Eye protection: Goggles or a face shield to shield the eyes from splashes or sprays.
- Shoe covers: Protect footwear from contamination.

Primary cell culture: In vitro culture of cells freshly obtained from the tissue is called primary cell culture. Examples, are primary mammary epithelial cells, primary hepatocyte culture

Sub culturing: is the process of removal of the medium from sub-confluent flask/dish and transfer of dissocated cells from a previous culture into fresh flask/dish containing growth medium upon trypsinization. By providing fresh nutrients and space, sub-culturing allows cells to continue growing and dividing for extended periods.

Suspension cell culture: cells that float and grow in the culture medium

$TCID_{50}$ is an endpoint dilution assay utilised to quantify the infectious virus titre.

Transfection is the intentional introduction of naked or purified nucleic acids into eukaryotic cells.

Transformed cell line is a cell line that has gained the ability to grow indefinitely due to the insertion of viral gene components into its genome.

Trypan Blue: Trypan blue is an azo dye used in biological science as a vital stain to color dead cells blue.

Viral transport medium is a solution that preserves virus samples for later analysis in a laboratory after collection.

Virus isolation: Virus isolation is a crucial testing method used to aid in diagnosing viral infections, including emerging, re-emerging, and novel viral pathogens. Isolating and identifying viruses through cell culture techniques or using embryonated chicken eggs

Virus neutralization is the process by which antibodies decrease the replication of viruses by preventing attachment to the host cell, inhibiting penetration of the host cell membrane, or interfering with uncoating of the virus within the cell.